SITE DESIGN

H. Paul Wood, AIA

KAPLAN AEC
ARCHITECTURE

President: Roy Lipner
Vice-President of Product Development and Publishing: Evan M. Butterfield
Editorial Project Manager: Michael J. Scafuri
Production Coordinator: Daniel Frey
Creative Director: Lucy Jenkins

Published by Kaplan AEC Architecture
a division of Dearborn Financial Publishing, Inc.®
A Kaplan Professional Company
30 South Wacker Drive
Chicago, IL 60606-7481
(312) 836-4400
http://www.ALSonline.com

Printed in the United States of America.

04 05 06   10 9 8 7 6 5 4 3 2

CONTENTS

Welcome to Kaplan Architecture! Kaplan recently acquired Architectural License Seminars (ALS), the oldest and most respected provider of Architect Registration Examination (ARE) study products. Over 50% of the registered architects in practice have used ALS material to prepare for their exam. Kaplan, Inc. is the nation's leading provider of lifelong education, having served more than 30 million individuals over the past 60 years.

THE ARE

Kaplan AEC Architecture provides the only complete, centralized source for all nine divisions of the ARE. All 50 states, 5 territories, and participating Canadian provinces offer the uniform NCARB Architect Registration Examination (ARE). This exam consists of nine divisions:

- Pre-Design
- General Structures
- Lateral Forces
- Mechanical & Electrical Systems
- Building Design/Materials & Methods
- Construction Documents & Services
- Site Planning
- Building Planning
- Building Technology

The Site Planning, Building Planning, and Building Technology divisions require graphic solutions to vignette graphic problems, while the other six divisions consist of multiple-choice questions. The exams are all administered by computer. Candidates must pass all divisions of the ARE in order to become a registered architect. Those who do not pass a division of the exam may retake it after six months. For further details on the ARE, please visit the NCARB Web site at www.ncarb.org.

Kaplan AEC Architecture provides a variety of study material to help you prepare for the exam: self study courses, computer mock exams, paper mock exams, question & answer handbooks, video workshops, and a CD-ROM-based flash card system. All of our products can be ordered online at www.ALSonline.com, or by calling (800) 420-1429.

The ARE is not an easy exam! Although we cannot guarantee a passing grade, we **can** guarantee our material will better prepare you for the ARE.

Good luck on your examination, and in your professional career.

ACKNOWLEDGMENTS

Kaplan AEC is grateful for the cooperation of many experts, professionals, organizations, and associations in the creation of this book. The following images are reproduced with the permission of the copyright holders.

Title	*Page*
Standard U.S.D.A Soil Classification Triangle	29
Unified Soil Classification System	30

Reprinted from the USDA. All rights reserved.

Table 18-1-A—Allowable Foundation and Lateral Pressure	33

Reprinted from *Uniform Building Code,* copyright International Conference of Building Officials, Whittier, California, 1997. All rights reserved.

This course of self-study is a condensed and simplified review of Site Design. It was developed as a study aid for the previous Site Design-Written test, but the information presented will be helpful to candidates preparing for the current Pre-Design or Site Planning test. Courses covering the other divisions of the ARE are also available, all of which have been prepared by the professional staff of ALS.

We advise that you set aside a definite amount of study time each week, just as if you were taking a lecture course, and carefully read all of the material. At the end of each lesson you will find a short review quiz. The quiz questions are intended to be straightforward and objective, and the answers and explanations are included to permit self-checking.

Following the last lesson, there is a Final Examination, which has been designed to simulate the style of the actual exam. Complete answers and explanations will be found on the pages following the examination, to permit self-grading.

You are now ready to begin your course in *Site Design*. We wish you success on your examination.

CLIMATE

INTRODUCTION

Background

For thousands of years, man has intuitively designed structures to be in harmony with nature, particularly with the climate. Primitive people understood how to build a shelter which would allow desirable natural elements, such as cooling summer breezes, to enter, while excluding the harsh winter winds. They knew how to provide wall openings to capture the warmth of the low winter sun, and they discovered that certain materials would retain the sun's heat and later release it to provide warmth after sundown.

In the past 50 years or so, as technology has advanced, we have become increasingly emancipated from climate. We work and live in buildings which are mechanically heated and cooled and which provide shelter against even the harshest climate. As a result, a certain

OVERHANG BLOCKS HIGH SUMMER SUN

THICK ADOBE WALLS RETAIN SOLAR HEAT & RELEASE IT AFTER SUNDOWN

SMALL OPENINGS REDUCE HEAT GAIN.

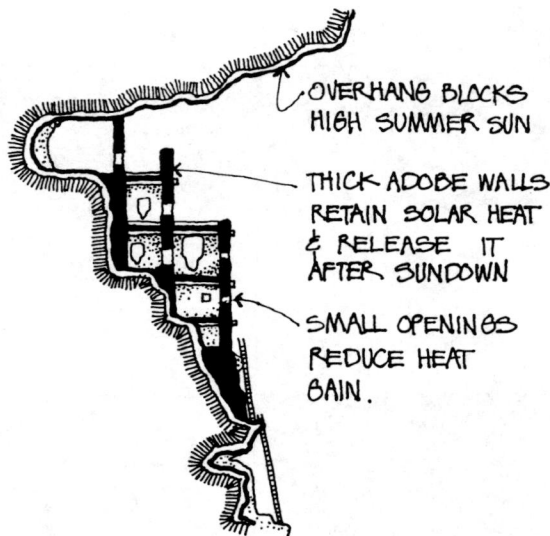

SECTION THROUGH INDIAN CLIFF DWELLING

uniformity of design has emerged: structures in the warm, humid climate of Miami often look exactly like those in Denver or New York, despite the vastly different climatic conditions. Also, buildings often have identical north, south, east, and west elevations, with total disregard for the different effects of sun and wind on the four sides. But energy shortages and the high cost of fossil fuels are changing all that. We can no longer afford to ignore nature; we must return to basics and relearn how to design sites and buildings that respond to the environmental forces that surround them.

Definitions

What is climate? And how does it differ from weather? The dictionary defines climate as "the generally prevailing weather conditions of a region throughout the year, averaged over a series of years." Weather, on the other hand, is the state of the atmosphere—temperature,

humidity, etc.—at a particular time. So weather is the momentary condition, while climate is the typical weather pattern at a given place, averaged over a period of time.

Climate consists of two parts: the general climate of a region, called the *macroclimate*, and its local modifications caused by the unique features of a site, known as the *microclimate*. Climate—macro or micro—consists of five major elements: sun, wind, temperature, humidity, and precipitation.

The macroclimate of an area is relatively constant. Despite the occasional rain in arid areas, or drought in wet areas, every place on earth has an inherent climatic pattern that tends to repeat itself over and over again. And this pattern is familiar to the residents of the area; people in Los Angeles, for example, know that they can expect some rain during the winter months, but virtually none at all between April and September.

What can we do about the macroclimate? Unfortunately, not much; our technology allows us to report the weather, predict it with varying degrees of accuracy, and shelter ourselves from it, but we remain unable to control or change it.

Microclimate, on the other hand, can be modified: we can plant trees, alter slopes, or create a lake or pool, and thus substantially change a site's microclimate.

Gathering Information

Before proceeding with any site design, the architect must determine what climatic information is necessary, and then obtain this data. He or she is often more interested in extremes than in average conditions. Information on the macroclimate is available from the National Weather Service and is relatively easy to get.

Microclimatic data, however, is not always readily available. The architect may visit the site a number of times to get familiar with its look, feel, smell, and sound. Studying the vegetation and the way natural and man-made features have weathered may provide clues about the microclimate. Talking to neighbors is also useful, since they are often the best source of information. After the architect has sifted and analyzed all this climatic data, he or she will be able to make basic site planning decisions—how to orient buildings on the site, how to protect them from the sun and wind, what modifications to the microclimate are appropriate, and how best to effect those changes.

MACROCLIMATE

Latitude

The macroclimate of an area is determined by a number of factors, chief among which is the amount of solar energy that it receives from the sun. This, in turn, depends on its latitude, or distance from the equator; in the Northern Hemisphere, the more northerly latitudes are further from the equator and therefore receive less solar energy.

Mountains and Oceans

But latitude is not the only factor shaping climate; mountains and oceans also play a part. As the elevation increases, the temperature decreases—about 3°F for every 1,000 feet—because the thinner air at higher altitudes cannot hold as much heat. This change of temperature is often experienced by visitors to Palm Springs, California who ride the Palm Springs Tramway. In 15 minutes, they are transported from the hot, arid desert floor to the alpine climate of Mt. San Jacinto at elevation 8,500 feet. The changes in temperature and vegetation during this brief ride are dramatic.

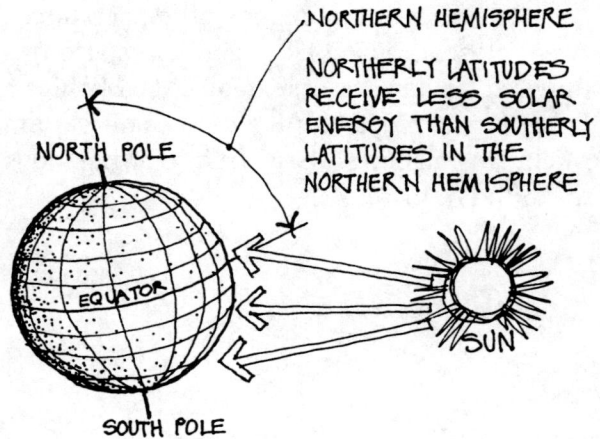

NORTHERN HEMISPHERE

NORTHERLY LATITUDES RECEIVE LESS SOLAR ENERGY THAN SOUTHERLY LATITUDES IN THE NORTHERN HEMISPHERE

NORTH POLE

EQUATOR

SUN

SOUTH POLE

SOLAR ENERGY DETERMINED BY LATITUDE

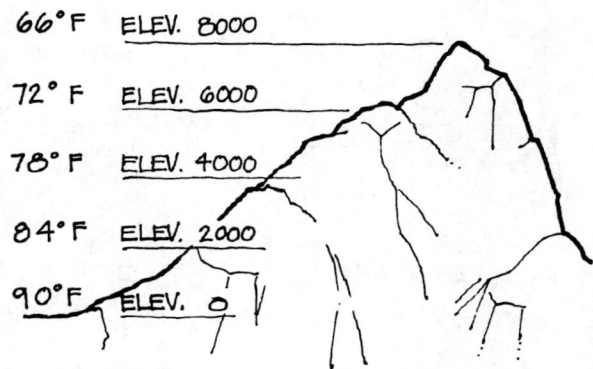

66°F ELEV. 8000

72°F ELEV. 6000

78°F ELEV. 4000

84°F ELEV. 2000

90°F ELEV. 0

TEMPERATURE CHANGES WITH ELEVATION

Since water heats up and cools down more slowly than land, the effect of bodies of water on nearby land areas is to moderate the climate by reducing temperature extremes; in the daytime, cool breezes from the sea move inland to replace rising warm air. At night, the flow reverses.

This moderating influence is most pronounced near the ocean; hence, coastal areas and islands tend to have a relatively uniform and moderate climate, while inland areas generally have wide daily and seasonal variations of temperature. Areas close to inland lakes and rivers also feel this effect, but to a lesser extent: Lake Michigan, for example, moderates the climate of western Michigan, enabling a large fruit-growing industry to thrive there.

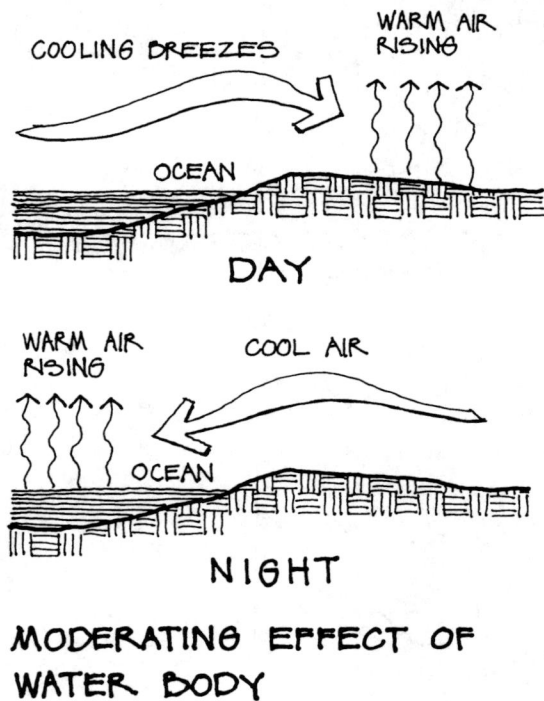

MODERATING EFFECT OF WATER BODY

Climate is also influenced by ocean currents; the Gulf Stream warms Norway, while the Labrador Current from the polar seas chills Newfoundland.

At mountain areas, prevailing winds drive the air up the windward slope, where it cools and often drops its moisture in the form of rain. The windward slopes therefore tend to be cool, humid, and vegetated, while the leeward slopes are generally warm and dry. An example of this effect is found in the state of Washington, where the rainfall on the westerly (windward) side of the Olympic Mountains increases dramatically as one moves higher in elevation, to as much as 150 to 200 inches annually. On the leeward side of the

mountains, the amount of precipitation drops to only 15 inches. A hill can create a similar effect, but to a lesser degree. Sometimes, the pattern of precipitation on a hill is the reverse of that on a mountain: rain on the windward side may be carried over the hill by the wind and fall on the leeward side, rather than on the windward side.

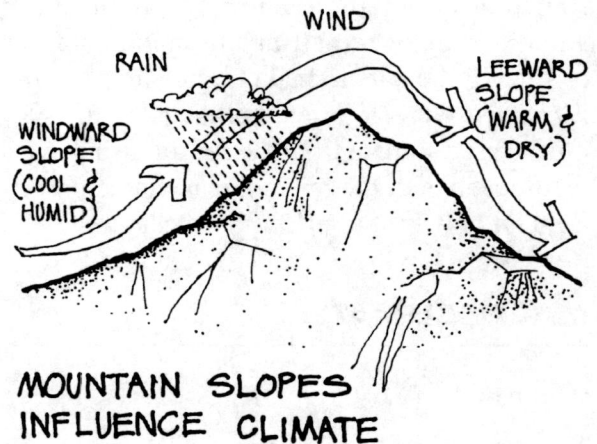

MOUNTAIN SLOPES INFLUENCE CLIMATE

Cloudiness is also a determinant of climate; clouds act as a blanket, making cloudy days cooler than clear days and cloudy nights warmer than clear nights.

CLIMATE ZONES

A number of systems have been devised to classify climatic zones within the United States. One simple scheme divides the country into four zones: cool, temperate, hot-arid, and hot-humid, as shown in the map above.

The boundary between zones is not exact—each zone merges gradually into the next one. And the climate characteristics within each zone are not uniform. The four-zone classification is simply a broad, general system which we use to help us understand the varying macroclimatic conditions in the country. Following is a brief description of the general climate in each zone.

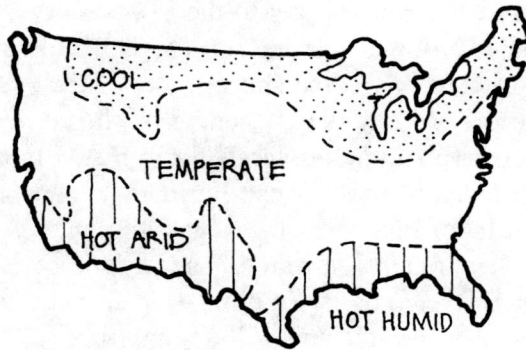

CLIMATE ZONES

Cool Zone

This zone is characterized by very cold, long winters, strong winds, and deep snow and frost. Consequently, site planning in this zone often involves controlling the wind, maximizing the winter sun, grouping activity areas to minimize outdoor travel time, and avoiding local cold air pockets.

Temperate Zone

This, the largest and most populous of the four regions, experiences great climatic variety. There are pronounced seasonal changes, ranging from warm to hot in the summer, cool to cold in the winter, and moderate in the spring and fall. As the song goes, "days may be cloudy or sunny," humid or dry, rainy or snowy. There is an ample supply of water, with many rivers and lakes, and a rich diversity of vegetation. Some site planning considerations include blocking cold winter winds while admitting cool summer breezes, maximizing shade in the summer, providing for extreme conditions of high wind, flooding, and snow, and incorporating natural waterways and indigenous vegetation into the landscape to preserve their unique ecological and aesthetic values

Hot-Arid Zone

The vast southwest, extending from Texas, through New Mexico, Arizona, southern Nevada, and into southern California, is a desertlike region characterized by clear skies, dry air, long periods of overheating, and large daily temperature variation. Days are hot, nights often cold, rainfall is minimal, and vegetation is sparse. The site planner uses shading and screening to provide relief from the heat and glare of the sun, preserves natural plant materials while adding compact planted spaces, and maximizes humidity and summer air movement. Flood plains, washes, and arroyos are avoided as sites for development because of occasional flash floods.

Hot-Humid Zone

The southeastern part of the country experiences high, relatively constant temperatures and humidity, variable winds which occasionally reach hurricane force, and torrential rain at times. Sites are designed to provide shade and air movement, while protecting against rains, flooding, and strong winds.

THE SUN

The sun, the source of all terrestrial energy, is the single most important natural element to consider in site and building design. The sun shapes all weather phenomena: solar energy causes air movement in the form of winds, the water cycle, which brings clouds and rain, the ocean currents, and glacial movements. We should therefore understand the path of the sun and the nature and intensity of its radiation.

The Sun's Path

In the Northern Hemisphere, the sun rises in the east, moves across the southerly sky, and sets in the west. Its path varies each day, but its position is entirely predictable and can be described at any time by its azimuth and its altitude. *Azimuth* is the angle through which one must turn, measured clockwise from south, in order to face the sun. *Altitude* is the angle between the horizon and the position of the sun above the horizon. The altitude is zero at sunrise and sunset, since the sun is then at the horizon. If the sun is directly overhead, the altitude is 90°.

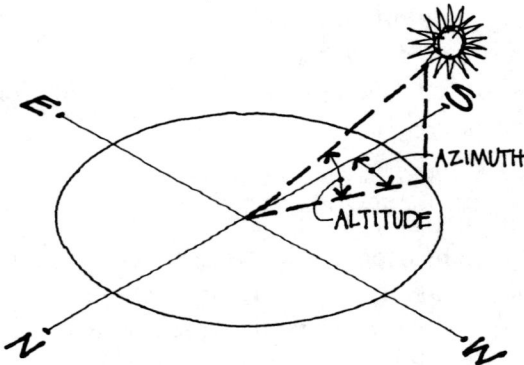

ALTITUDE AND AZIMUTH

A number of methods to graphically depict the sun's path are commonly used. One such method is the sun chart, shown below for 40° north latitude. Every place on earth whose latitude is 40° north has the same sun chart.

The chart is essentially a map of the sky, with the concentric circles representing altitude and the radial lines representing azimuth. If you were to lie flat on the ground on the twenty-first day of each month, look straight up, and plot the path of the sun, you would arrive at the seven curved sun paths shown, identified by Roman numerals representing the months. The upper-most curve, labeled VI, is for June; the next curve, labeled V and VII, is for the months of May and July, which have the identical sun path;

and so on, until we get to the lowest curve, labeled XII, which represents the path of the sun on December 21, the shortest day of the year. The time of day is represented by the curved lines with Arabic numerals, from 5 AM in the east to 7 PM in the west. For dates other than the 21st of each month, we can interpolate. And for locations other than 40° north latitude, we use different charts. So now, we can accurately predict where the sun will be, any time of day or any day of the year.

We can see that in the winter, the sun is low and to the south. Therefore, the south side of a building will receive a great amount of solar radiation, while the east and west faces receive very little. Horizontal surfaces, such as roofs, also receive little solar energy. In the summer, the situation is reversed—the sun is higher and is oriented more to the east and west. Consequently, the south side receives less radiation, while the east and west sides and roof receive more. The north side receives little radiation summer or winter.

Solar Orientation

The major glass areas of a building should receive the maximum amount of solar radiation in the winter and the minimum amount in the summer and should therefore face south to southeast. East and west walls, since they receive most of their radiation in the summer, should have less wall exposure and glass, to avoid excessive heat gain in summer. The north wall, which receives the least amount of solar radiation summer or winter, should have the fewest window openings, to avoid excessive heat loss in winter.

Thus, considering climate only, a building should ideally have its major axis approximately east-west, with maximum glass in the south-to southeast-facing wall, moderate glass in the east

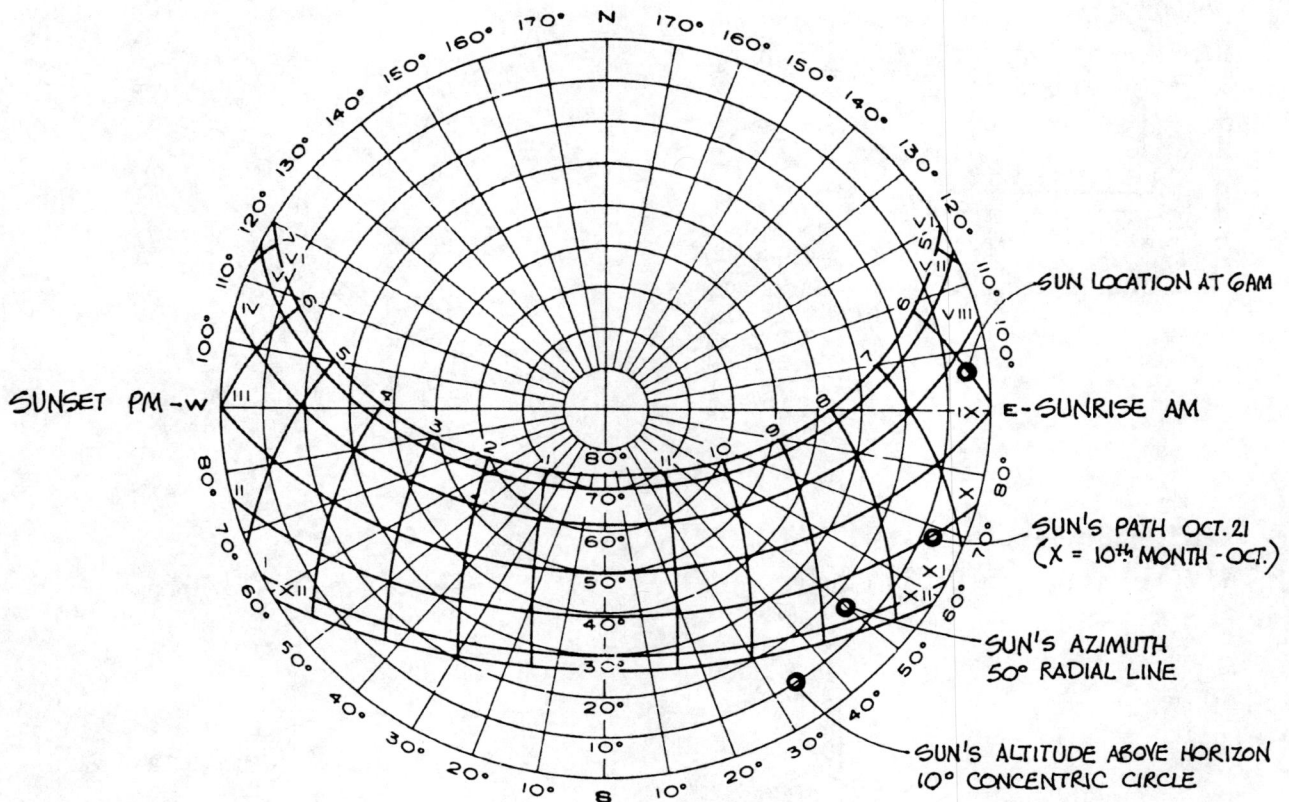

Labels on the chart:

SUNSET PM — W

SUN LOCATION AT 6AM

E — SUNRISE AM

SUN'S PATH OCT. 21
(X = 10th MONTH - OCT.)

SUN'S AZIMUTH
50° RADIAL LINE

SUN'S ALTITUDE ABOVE HORIZON
10° CONCENTRIC CIRCLE

SUN CHART FOR 40° NORTH LATITUDE

and west walls, and minimum glass in the north wall. The south wall should be shaded by an overhang to block the high summer sun, while allowing the low winter sun to enter.

The discussion above applies mainly to cool and temperate regions. In warm climates, some modifications may be in order. For example, major glass areas may face north, to minimize heat gain.

Solar Radiation

The sun's energy travels through space in the form of electromagnetic radiation. Much of this solar energy never reaches us, but is absorbed by material in the atmosphere or reflected back into space by clouds. The radiation which does reach us is in two forms: direct radiation from the sun and diffuse radiation from all parts of the sky.

The amount of solar radiation received by a surface is determined by the angle that the sun's rays make with the surface; a surface perpendicular to the sun's rays will receive the maximum amount of energy, and as the sun's rays become less perpendicular to the surface, less energy is received. Thus, since the sun moves in a southerly path, south-facing slopes receive more energy than level or north-facing slopes.

HIGH SUMMER SUN WARMS THE ROOF AND EAST & WEST WALLS.

LOW WINTER SUN WARMS SOUTH WALL

OVERHANG ALLOWS WINTER SUN TO ENTER BUT BLOCKS OUT SUMMER SUN.

SOUTH

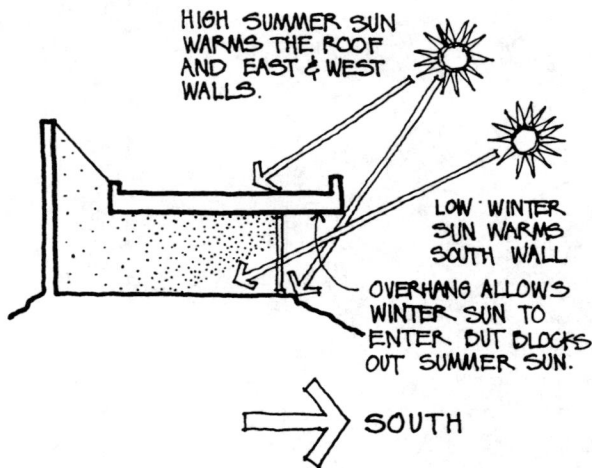

THE SUN'S LOCATION VARIES WITH THE SEASONS

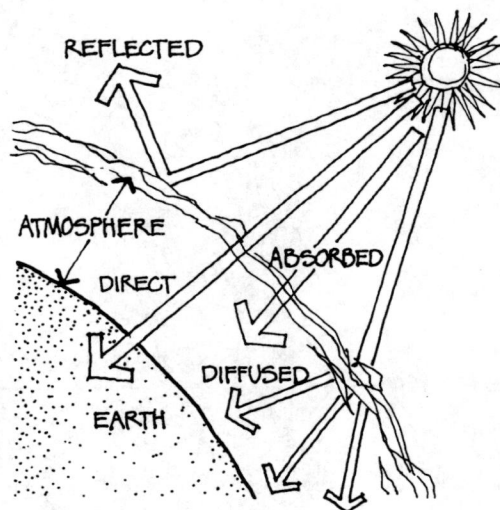

REFLECTED

ATMOSPHERE

DIRECT

ABSORBED

DIFFUSED

EARTH

SOLAR RADIATION

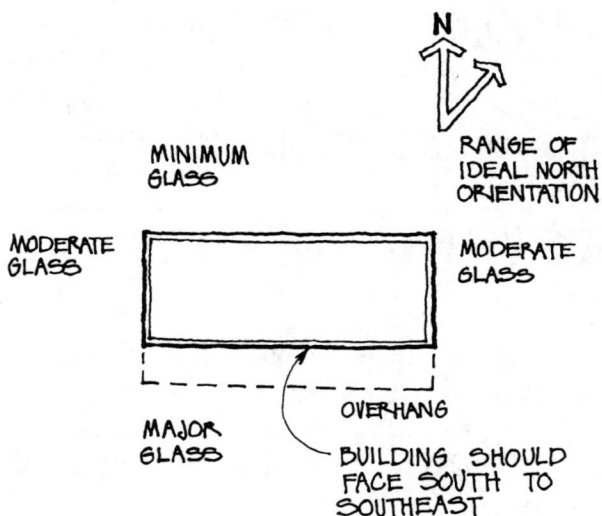

N

RANGE OF IDEAL NORTH ORIENTATION

MINIMUM GLASS

MODERATE GLASS

MODERATE GLASS

MAJOR GLASS

OVERHANG

BUILDING SHOULD FACE SOUTH TO SOUTHEAST

IDEAL ORIENTATION

SHALLOW ANGLE

PERPENDICULAR

SOUTH-FACING SLOPE RECEIVES MORE ENERGY THAN LEVEL SITE

SOUTH

NORTH

SOLAR ENERGY DETERMINED BY ANGLE OF SURFACE

The arrival of spring on a south-facing slope may be weeks ahead of its arrival on a level site.

By referring to the sun chart on page 7, we see that north-facing slopes receive no direct sun exposure, except in the early morning and late afternoon of summer. East-facing slopes receive direct sun only in the morning, particularly in

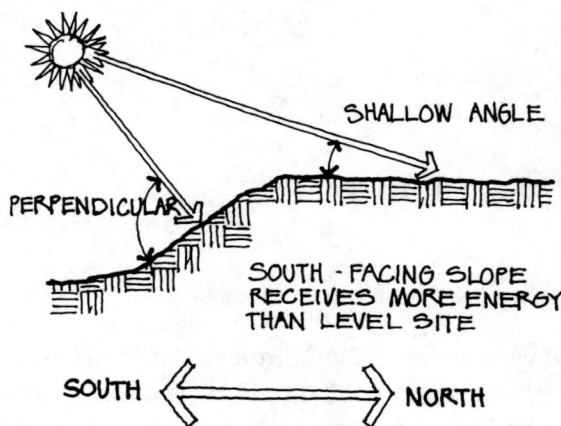

the summer, while west-facing slopes are exposed to the afternoon sun, which can be hot and penetrating in the summer.

The amount of solar radiation reaching the earth's surface is also determined by the length of travel through the atmosphere. During the day, when the sun is high, radiation travels through the least amount of atmosphere and

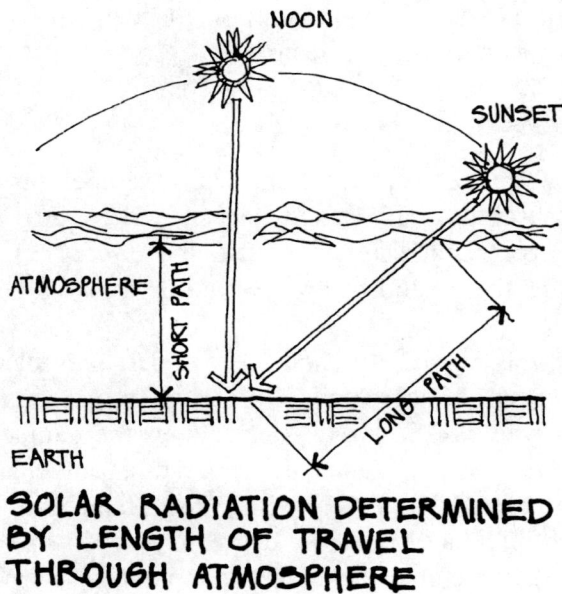

SOLAR RADIATION DETERMINED BY LENGTH OF TRAVEL THROUGH ATMOSPHERE

therefore the earth receives maximum radiation. As the sun drops lower in the sky, towards sunset, the length of travel through the atmosphere increases and less energy reaches us. Similarly, at high altitudes, there is less atmosphere and therefore more solar radiation.

The Seasons

How do we explain the seasons? As the earth orbits the sun, it rotates once a day about its own north-south axis. If this axis were perpendicular to the plane of the earth's orbit around the sun, there would be no seasons: the sun's path across the sky would be the same every day of the year. But the earth's axis is tilted 23-1/2° from a line perpendicular to the plane of the earth's orbit around the sun. As a result, the Northern Hemisphere is slanted toward the sun in the summer and away from the sun in the winter. In the summer, the sun's rays are close to perpendicular to the earth's surface and the length of atmosphere through which solar radiation passes is minimum. Therefore, a maximum amount of solar radiation reaches the earth. In the winter, the situation is reversed and the solar energy reaching us is minimum.

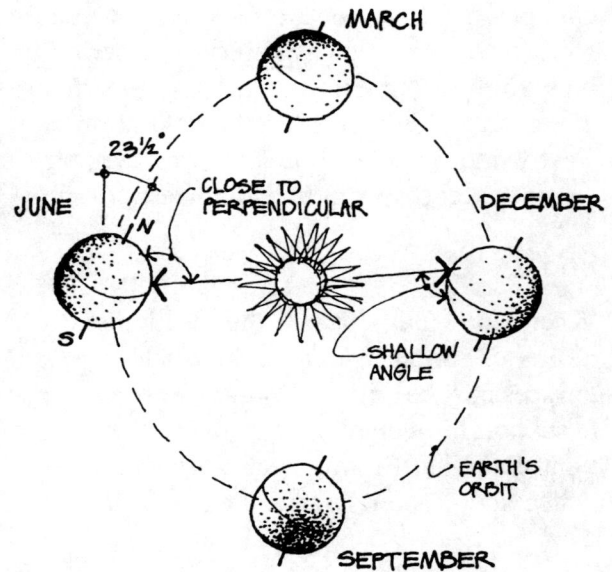

THE EARTH'S TILT CREATES THE SEASONS

OTHER CLIMATIC FACTORS

Wind

Climatic factors other than the sun must also be considered in site design. Winds increase heat loss from both buildings and people. This may be desirable in the summer months, but detrimental in the winter. Winds also blow away the insulating layer of still air that normally surrounds a building, thus increasing both heating and cooling loads.

Cold winter winds in the United States typically come from the north and west. Therefore, entrances and windows on the north and west sides of a building should be minimized, or avoided altogether. In addition, trees or other windbreaks may be used to block the cold winds.

Summer breezes, on the other hand, generally come from the southwest, and may be channeled to cool both interior and exterior spaces.

Outdoor living spaces are most often located south or southeast of dwellings or other buildings. In that location, they receive maximum winter sun, shielding from the cold north and west winter winds, while admitting cooling summer breezes from the southwest.

Snow

Where snowfall is heavy, the location and design of entrances and outdoor balconies and terraces may be critical. Roofs must be designed to support the weight of the snow, and the reflection of sunlight from the snow creates glare, which should be recognized and considered.

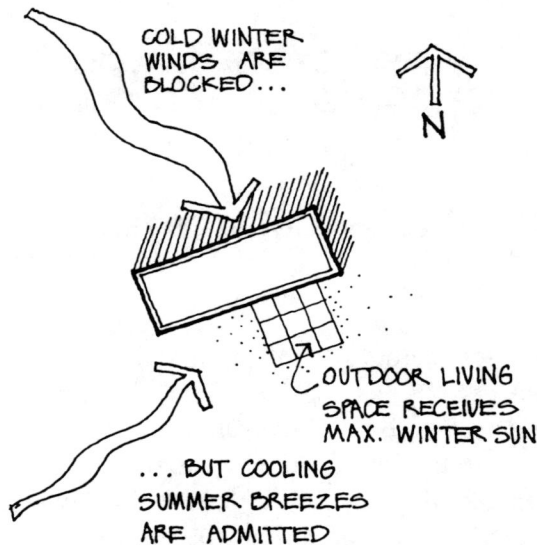

OUTDOOR SPACE SOUTHEAST OF BUILDING

Humidity

Humidity is generally described as relative humidity—the ratio of the actual amount of moisture in the air to the maximum amount of moisture the air could hold at a given temperature. At temperatures above 65°F, high relative humidities create discomfort, and relief in the

form of natural ventilation or mechanical cooling is necessary to restore comfort.

Degree Days

The range of summer and winter temperatures is used not only to size heating and cooling systems, but also as a guide to the outdoor activities that would be appropriate on a given site.

In this regard, it is helpful to understand the concept of degree days. The number of heating degree days in a day is defined as the number of degrees that the mean temperature for the day is below 65°F. A day with a mean temperature of 50°F has 15 heating degree days (65 − 50 = 15), while one whose mean temperature is 65°F or higher has no degree days. The number of degree days in a year is simply the total of the degree days for every day of the year, and this gives us a good indication of the energy required to maintain a constant indoor temperature for buildings in that area. One can see from the map below that the number of degree days in this country varies widely, from zero in south Florida to 10,000 in northern Minnesota, and even higher in Alaska.

HUMAN COMFORT

Climate is a factor of considerable importance in site and building design, since it affects people both physically and emotionally. The architect's goal is to create an environment, both indoors and outdoors, which will provide complete human comfort—a feeling of total physical and psychological well-being.

Four climatic factors affect human comfort, and all of them must be considered simultaneously in order to create a comfortable indoor or outdoor environment. These factors are: 1) temperature, 2) humidity, 3) radiation, and 4) air movement.

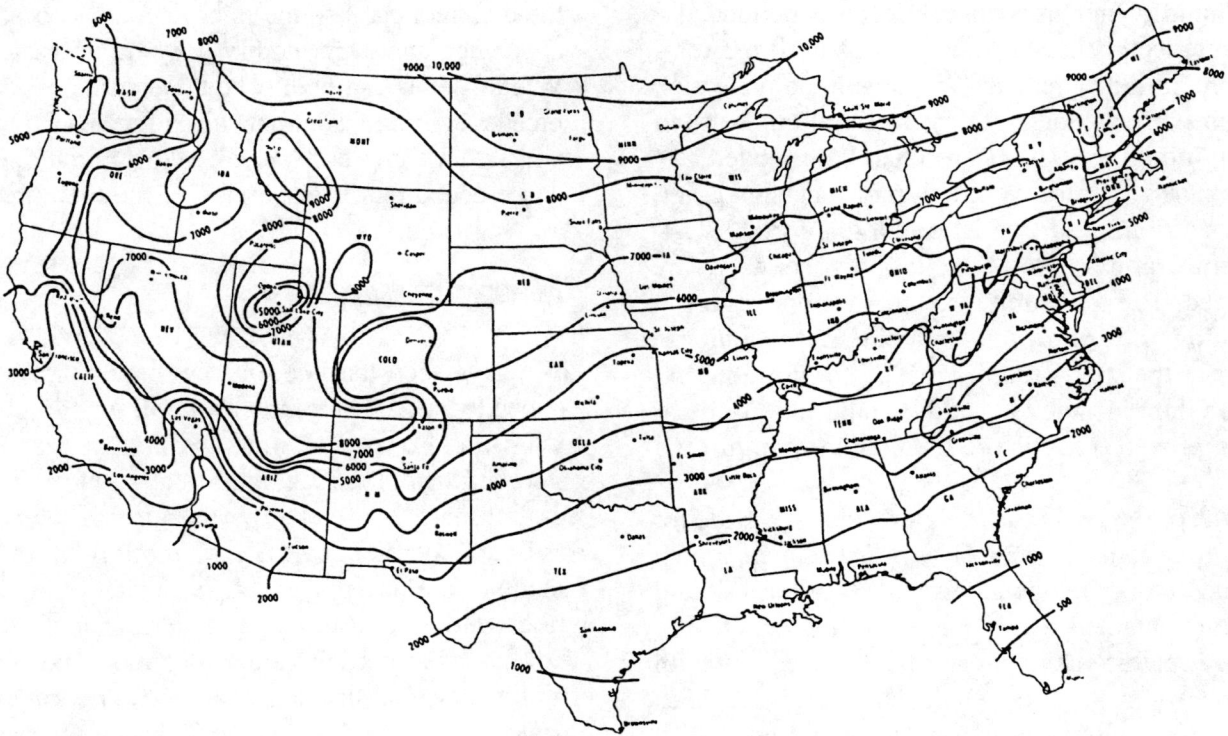

NUMBER OF DEGREE DAYS PER YEAR

Body's Thermal Processes

In order to effectively design for human comfort, we must first have some understanding of the body's thermal processes. The body is a machine which continuously produces heat, the amount of which varies according to the individual's age, size, health, and physical activity. When sleeping, the average person produces about 300 BTU per hour; at maximum exertion, this increases to over 4,000 BTU per hour. The body dissipates this heat to the environment principally by radiation and convection, with lesser amounts given up by evaporation and conduction. The body thus maintains a constant internal temperature of 98.6°F, and any significant departure from this temperature causes discomfort. If the body's heat gain exceeds its heat loss, we are uncomfortable and we perspire;

conversely, if heat loss is greater than heat gain, our body temperature drops and we shiver.

Comfort Zone

The comfort zone consists of that combination of temperature and humidity in which the average person feels comfortable. It is at best an imprecise approximation, since it varies with many factors, including climate, culture, and the individual's sex, age, clothing, and type of activity. In this country, the comfort zone is roughly between 65°F and 75°F and between 20 percent and 75 percent relative humidity, as shown in the chart on the following page.

Any combination of temperature and humidity within the unshaded area will be comfortable, while all other combinations will be uncomfortable. You can see that as the humidity increases, discomfort is experienced at increasingly lower temperatures until the relative humidity reaches 75 percent, above which discomfort results at any temperature. At such high humidity, moisture must be removed, either by natural ventilation or mechanical air conditioning, if one is to feel comfortable. At temperatures above the comfort zone, winds are needed to feel comfortable. At temperatures below the comfort zone, additional radiation from the sun is required to be comfortable. At low humidity levels, additional humidity must be provided to restore the feeling of comfort.

One of the goals of the architect is to provide for human comfort; as much as possible, then, indoor and outdoor temperatures and humidities should be maintained within the comfort zone. Of course, this cannot always be achieved, but with an awareness of the relationships among temperature, humidity, radiation, and air movement,

the architect can use the microclimate to best advantage and even modify it where necessary. Winter winds can be blocked and summer breezes admitted; summer sun can be shaded and winter sun admitted. With all of this in mind, we can proceed to our discussion of microclimate.

MICROCLIMATE

We have seen that the macroclimate is influenced by such factors as elevation, topography, winds, and bodies of water. Similarly, the microclimate, the local modification of the general climate, is affected by those same factors, only at a smaller scale. In addition, other elements come into play, such as vegetation, structures, and ground cover. And unlike the general climate, the microclimate can be modified. The architect's goal in manipulating microclimatic factors is always the same: to make the site more

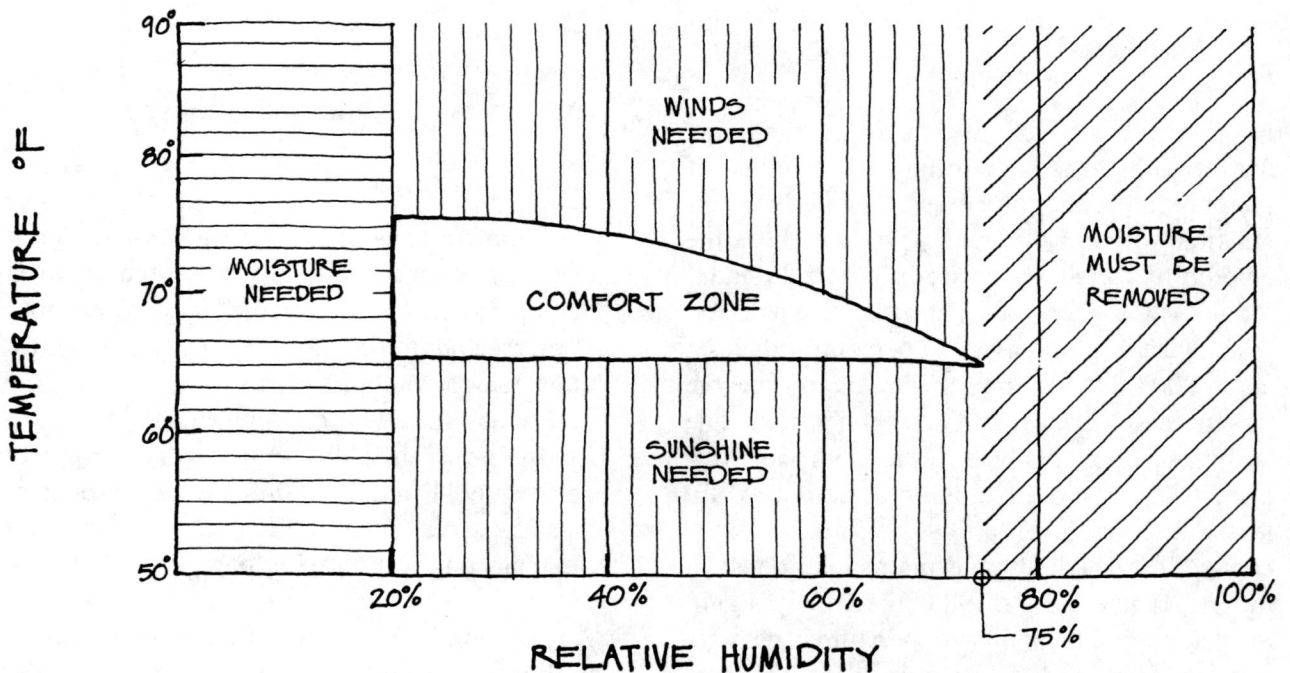

DIAGRAMMATIC BIOCLIMATIC CHART

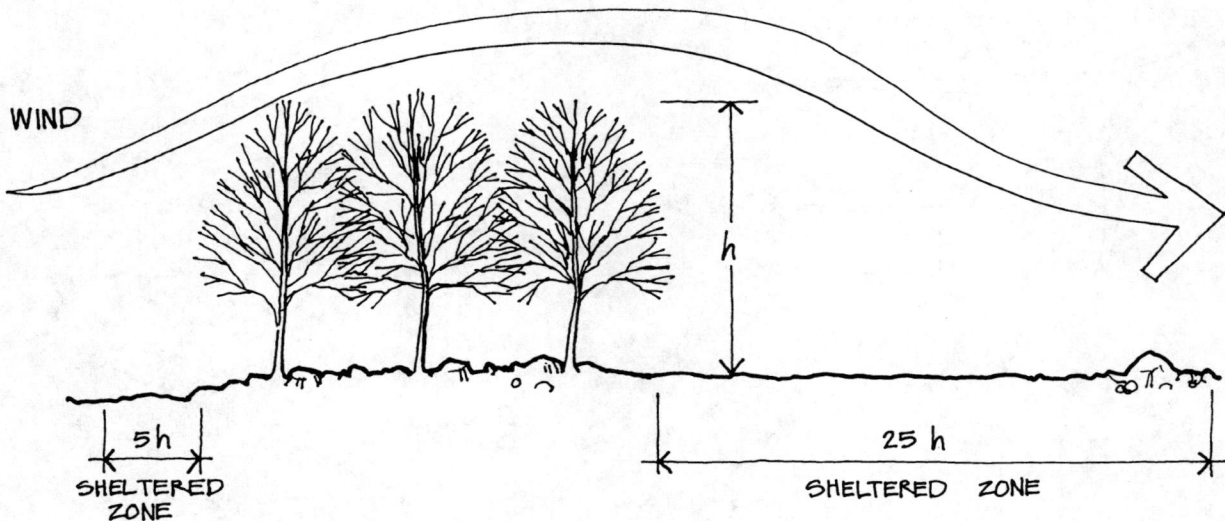

TREES PROVIDE PROTECTION FROM THE WIND

comfortable. Let's discuss some of the elements of microclimate in greater detail.

Water

A body of water moderates the microclimate. Therefore, the site planner will usually preserve any existing streams on a site. He or she may also introduce water in the form of a pool, a reflecting pool, a fountain, or a waterfall. The moderating effect—especially cooling in the summer—is not only physical, but also

psychological, through the sound of falling or running water.

Trees

Trees have a great effect on microclimate: they provide shade, they block the wind, they cool, humidify, and filter the air, and they help to prevent rapid, destructive runoff. Trees substantially reduce wind speeds and thus create sheltered zones on both the windward and leeward sides of a wooded area, the leeward zone being much greater than that on the windward side.

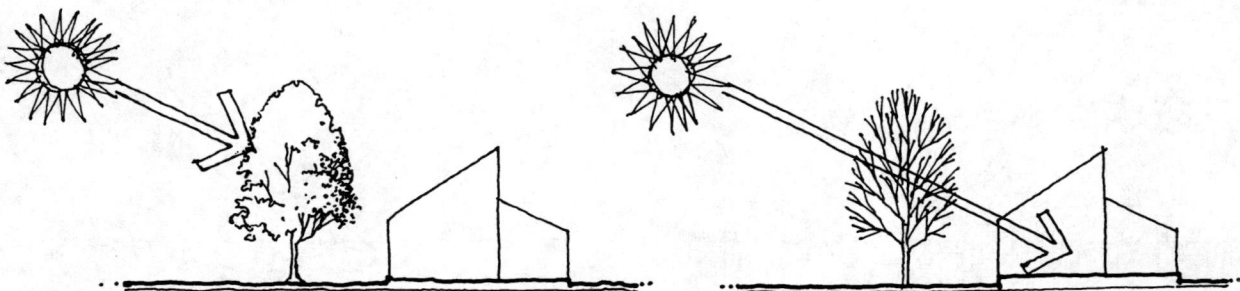

DECIDUOUS TREES BLOCK THE SUMMER SUN...BUT ALLOW WINTER SUN TO PENETRATE.

DECIDUOUS TREES

WIND

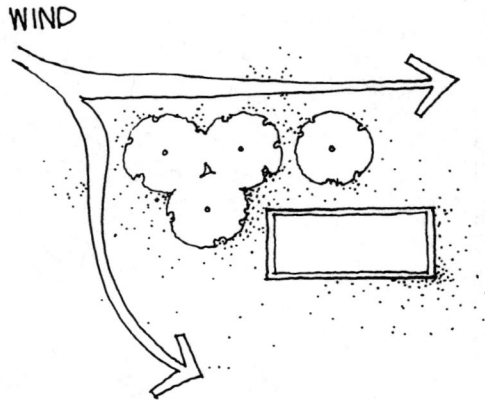

TREES CAN DEFLECT THE WIND ...

WIND

... OR CREATE A COLD AIR POCKET.

WINDS AFFECTED BY TREE LOCATIONS

Deciduous trees, such as maple and beech, are especially advantageous for blocking the sun, since they shade only the summer sun, but allow the winter sun to penetrate. Evergreen trees, including firs and pines, are effective in blocking winter winds, since they retain their foliage all year. For the same reason, they are often used to provide permanent privacy and screening. Properly placed trees can deflect or channel winds, but if they are not located correctly, they can create an undesirable pocket of cold air.

Albedo

When we speak of ground cover, we sometimes use the term "albedo," which essentially refers

to reflectivity. Natural ground cover, such as grass, has a lower value of albedo than a paved surface, and therefore reflects less heat. On a warm day, temperatures over grass may be 20° cooler than over pavement; grass also has less dust and glare and absorbs more sound than a paved surface. For these reasons, it is usually advisable to minimize paving immediately adjacent to a building, and introduce grass or other vegetation instead.

Hills

Cold air, being heavier than warm air, flows downhill. As a result, in hilly country, the coldest ground surface and air are found in the

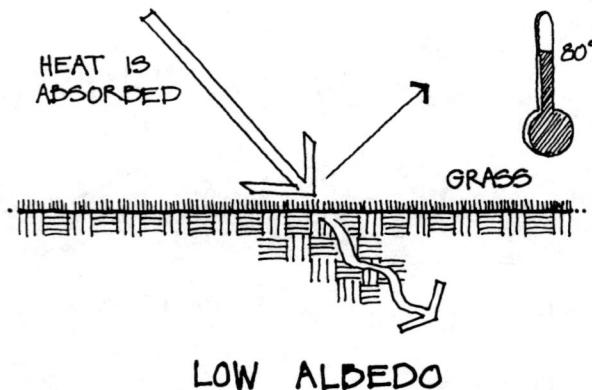

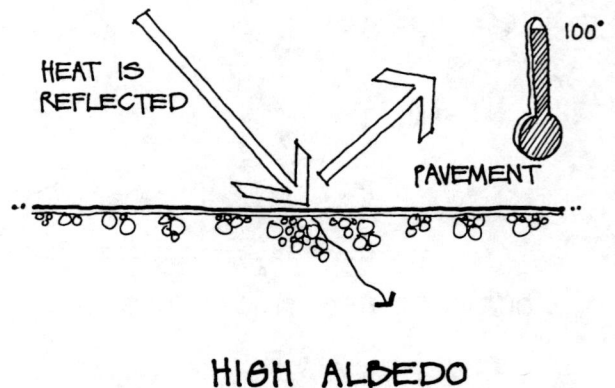

HEAT IS
ABSORBED

80°

GRASS

LOW ALBEDO

HEAT IS
REFLECTED

100°

PAVEMENT

HIGH ALBEDO

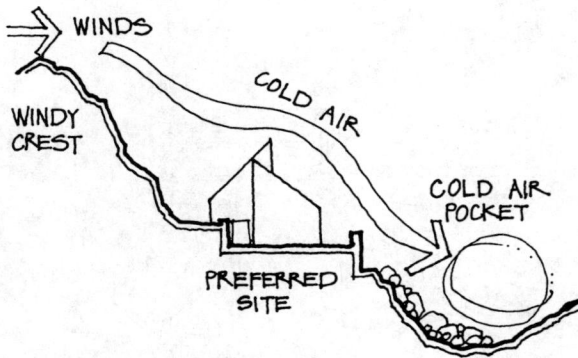

HILLS AFFECT MICROCLIMATE

valleys. Valley sites may therefore be desirable in warm climates, but undesirable in cooler climates. Thus, a site on a slope in a cooler climate is preferable to either the cold valley floor or the windy crest. And as we've already mentioned, south-facing slopes are warmer in the winter than level sites or sites facing in any other direction.

Landforms and Structures

Site elements other than trees can also block sun and wind; whether by design or otherwise, landforms, fences, and structures also shade the sun and deflect the wind. On the windward side of a hill, wind speeds are highest near the crest, while the leeward slope has less turbulent winds. Near the bottom of the hill on the leeward side, the winds decrease to almost zero, creating a "wind shadow."

A high fence may be used to effectively shade a building or outdoor space from the hot late afternoon sun of summer, when the sun is low, if it is located to the west and northwest of the building or space to be protected. Fences are much less effective as shading devices when the sun is high in the sky.

The variations are endless, depending on the climate, the height and orientation of blocking elements, and prevailing winds.

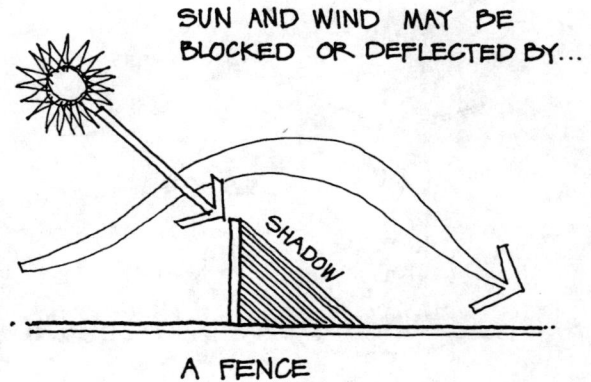

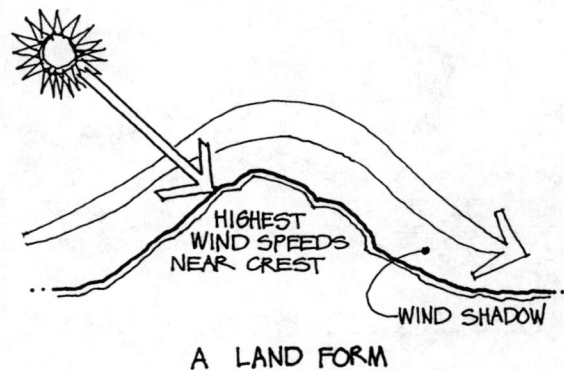

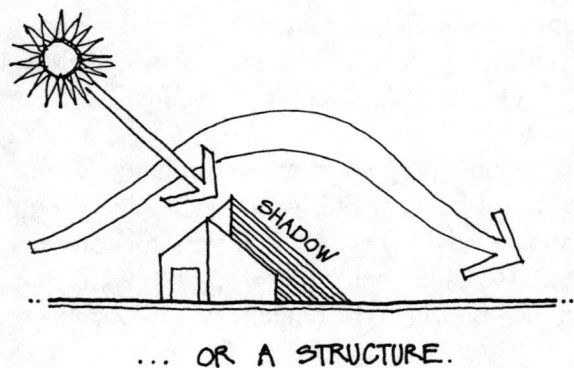

SUN AND WIND MAY BE BLOCKED OR DEFLECTED BY...

A FENCE

A LAND FORM

... OR A STRUCTURE.

MICROCLIMATIC ELEMENTS BLOCK THE SUN & WIND

Other Factors

Structures may be built into the ground for natural insulation and to minimize exposure to winds.

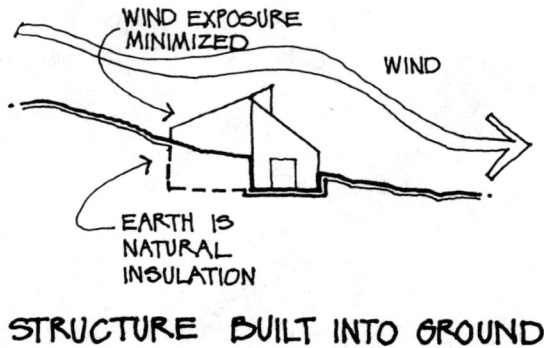

STRUCTURE BUILT INTO GROUND

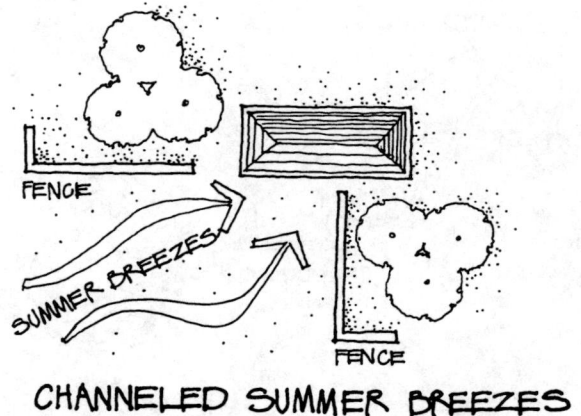

CHANNELED SUMMER BREEZES

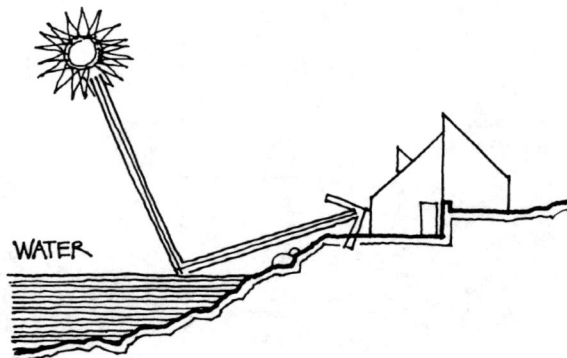

GLARE MUST BE CONSIDERED

Glare from adjacent water or snow can be an unpleasant factor to contend with, and the architect may want to minimize west-facing windows and outdoor activity areas in such cases.

Just as the site planner tries to minimize the effects of harsh winter winds, he or she should also take advantage of cooling summer breezes. Walls or hedges can be carefully placed to channel the breezes to cool indoor or outdoor spaces.

Pedestrian circulation routes should be shaded by trees or arcades, and deciduous trees should be located south of such walks to allow the winter sun to penetrate.

Urban Microclimate

The urban microclimate differs considerably from that of the suburbs or the rural countryside.

The agglomeration of streets, parking areas, and structures that characterizes our cities, along with the elimination of natural ground cover and the emission of heat, creates a microclimate which is warmer and drier than rural areas, with more rain, clouds, and fog. There is also more noise, air pollution, and glare. Urban structures also block wind and sunshine. As we learn more about microclimate, perhaps we will be able to improve the city's climate, by controlling air pollution, adding parks, plazas, and fountains, and using mechanical heating and cooling devices where appropriate.

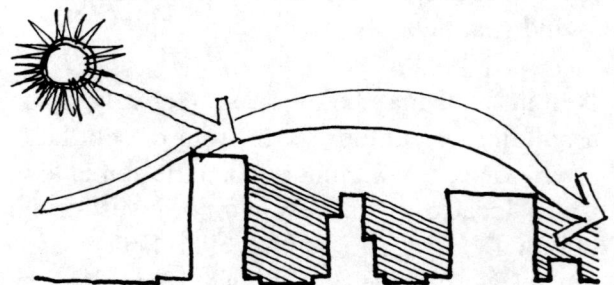

URBAN STRUCTURES BLOCK WIND AND SUNSHINE

SOLAR ENERGY

Background

Buildings consume tremendous amounts of energy, for space heating and cooling, lighting, power, and hot water heating. Almost all of this energy is produced by the burning of fossil fuels (coal, oil, and gas), either on-site or off-site to generate electricity. But these fuels, especially oil and gas, have become expensive, their supply is limited and somewhat controlled by the oil-exporting countries, and they are polluting and non-renewable. It is clear, therefore, that we must seriously consider the sun as an alternative source of energy. It is not our purpose here to discuss the details of solar energy, but rather to present a brief overview of the subject.

Technology

The basic technology for using solar energy to heat and cool buildings already exists, but it must be further refined and developed before solar energy can gain widespread acceptance and use. While it is theoretically possible for a building to be heated and cooled 100 percent by solar radiation, a more feasible goal is 70 percent for solar space heating and 90 percent for solar hot water heating. Therefore, a solar space heating or hot water heating system must be supplemented by an auxiliary system, which may use either fossil fuels or some other energy source. Solar cooling, although technically feasible, is not yet as efficient as solar heating.

Flat Plate Collectors

The most common device used to capture the sun's energy is the flat plate collector, which can be located on the roof, an adjacent building, the ground, or the south wall. Collectors generally face south and are tilted to receive maximum solar radiation. The optimum tilt of a flat plate collector varies with the latitude and the use, as shown in the sketch below. Thus, at 40° north latitude, a collector used only for hot water heating would be tilted at 45° (latitude + 5°). At more southerly latitudes, since the sun is more directly overhead, the tilt would be less—that is, the collector would be closer to horizontal. Of course, collectors should not be shaded by trees, landforms, or other structures.

A number of other possibilities exist for the use of solar energy in buildings, including the generation of electricity directly from photovoltaic cells; however, such systems are not generally available or economical as yet.

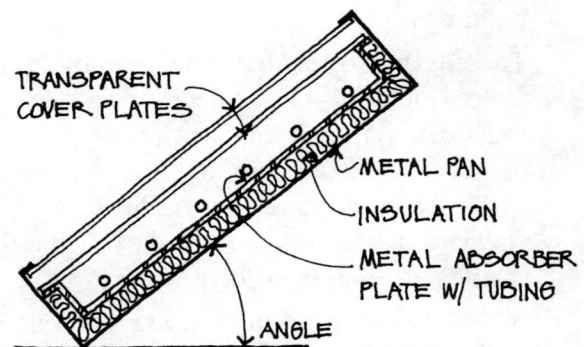

TRANSPARENT COVER PLATES
METAL PAN
INSULATION
METAL ABSORBER PLATE W/ TUBING
ANGLE

ANGLE = LATITUDE + 15° FOR HEATING
LATITUDE + 5° FOR HTG./COOLING
LATITUDE + 5° FOR HOT WATER
LATITUDE + 15° FOR HTG./HOT WATER

TYPICAL FLAT PLATE COLLECTOR

SUSTAINABLE SITE PLANNING AND DESIGN

Most architectural projects involve the understanding of the design within the context of the larger scale neighborhood, community, or urban context in which the project is placed.

If the building will be influenced by sustainable design principles, its context and site should be equally sensitive to environmental planning principles.

Sustainable design encourages a re-examination of the principles of planning to include a more environmentally sensitive approach. Whether it is called Smart Grow, sustainable design, or environmentally sensitive development practice, these planning approaches have several principles in common.

1. **Site Selection**

   The architect and planner may assist the client in developing the criteria for site selection that reflects the proposed environmental goals of the complex of buildings.

   The selection of a site is influenced by many factors including cost, adjacency to utilities, transportation, building type, zoning, and neighborhood compatibility. But, in addition to these factors, there are sustainable design standards that should be added to the matrix of site selection decisions:

   ■ *Adjacency to Public Transportation:*
   If possible, projects that allow residents or employees access to public transportation are preferred. Allowing the building occupants the option of traveling by public transit may decrease the parking requirements, increase the pool of potential employees and remove the stress and expense of commuting by car.

   ■ *Flood Plains:*
   In general, local and national governments hope to remove buildings from the level of the 100-year flood plain. This can be accomplished by either raising the building at least one foot above the 100-year elevation or locating the project entirely out of the 100-year flood plain.

This approach reduces the possibility of damage from flood waters, and possible damage to downstream structures hit by the overfilled capacity of the floodplain.

■ *Erosion, Fire, and Landslides:*
Some ecosystems are naturally prone to fire and erosion cycles. Areas such as high slope, chaparral ecologies are prone to fires and mud slides. Building in such zones is hazardous and damaging to the ecosystem and should be avoided.

■ *Sites with high slope or agricultural use:*
Sites with high slopes are difficult building sites and may disturb ecosystems, which may lead to erosion and topsoil loss. Similarly, sites with fertile topsoil conditions—prime agricultural sites—should be preserved for crops, wildlife, and plant material, not building development.

■ *Solar orientation, wind patterns:*
Orienting the building with the long axis generally east west and fenestration primarily facing south may have a strong impact on solar harvesting potential. In addition, protecting the building with earth forms and tree lines may reduce the heat loss in the winter and diminish summer heat gain.

■ *Landscape Site Conditions:*
The location of dense, coniferous trees on the elevation against the prevailing wind (usually west or northwest) may decrease heat loss due to infiltration and wind chill factor. Sites with deciduous shade trees can reduce summer solar gain if positioned properly on the south and west elevations of the buildings.

2. **Alternative Transportation**

Sites that are near facilities that allow several transportation options should be encouraged. Alternate transportation includes public transportation (trains, buses, and vans); bicycling amenities (bike paths, shelters, ramps, and overpasses); carpool opportunities that may also connect with mass transit; and provisions for alternate, more environmentally sensitive fuel options such as electricity or hydrogen.

3. **Reduce Site Disturbance**

   Site selection should conserve natural areas, and restore wildlife habitat and ecologically damaged areas. In some areas of the United States, less than 2 percent of the original vegetation remains. Natural areas provide a visual and physical barrier between high activity zones. Additionally, these natural areas are aesthetic and psychological refuges for humans and wildlife.

4. **Storm Water Management**

   Reduced disruption of natural water courses (rivers, streams, and natural drainage swales) may be achieved by:

   - Providing on-site infiltration of contaminants (especially petrochemicals) from entering the main waterways. Drainage designs that use swales filled with wetland vegetation is a natural filtration technique especially useful in parking and large grass areas.

   - Reducing impermeable surface and allowing local aquifer recharge instead of runoff to waterways.

   - Encouraging groundwater recharge.

5. **Ecologically Sensitive Landscaping**

   The selection of indigenous plant material, contouring the land, and proper positioning of shade trees may have a positive effect on the landscape appearance, maintenance cost, and ecological balance. The following are some basic sustainable landscape techniques:

   - Install indigenous plant material, which is usually less expensive, to ensure durability (being originally intended for that climate) and lower maintenance (usually less watering and fertilizer).

   - Locate shade trees and plants over dark surfaces to reduce the "heat island effect" of surfaces (such as parking lots, cars, walkways) that will otherwise absorb direct solar radiation and retransmit it to the atmosphere.

   - Replace lawns with natural grasses. Lawns require heavy maintenance including watering, fertilizer, and mowing. Sustainable design encourages indigenous plant material that is aesthetically compelling but far less ecologically disruptive.

   - In dry climates, encourage xeriscaping (plant materials adapted to dry and desert climates); encourage higher efficiency irrigation technologies including drip irrigation, rainwater recapture and gray water reuse. High efficiency irrigation uses less water because it supplies directly to the plant's root areas.

6. **Reduce Light Pollution**

   Lighting of site conditions, either the buildings or landscaping, should not transgress the property and not shine into the atmosphere. Such practice is wasteful and irritating to the inhabitants of surrounding properties. All site lighting should be directed downward to avoid "light pollution."

7. **Open Space Preservation**

   The quality of residential and commercial life benefits from opportunities to recreate and experience open-space areas. These parks, wild life refuges, easements, bike paths, wetlands, or play lots are amenities that are necessary for any development.

In addition to the aforementioned water management principles, the following are principles of design and planning that will help increase open-space preservation:

7.1 Promote in-fill development that is compact and contiguous to existing infrastructure and public transportation opportunities.

In-fill development may take advantage of already disturbed land without impinging on existing natural and agricultural land.

In certain cases, in-fill or redevelopment may take advantage of existing rather than new infrastructure.

7.2 Promote development that protects natural resources and provides buffers between natural and intensive use areas.

- First, identify the natural areas (wetlands, wildlife habitats, water bodies, or flood plains) in the community in which the design is planned.

- Second, the architect and planners should provide a design that protects and enhances the natural areas. The areas may be used partly for recreation, parks, natural habitats, and environmental education.

- Third, the design should provide natural buffers (such as woodlands and grasslands) between sensitive natural areas and areas of intense use (factories, commercial districts, housing). These buffers may be both visual, olfactory, and auditory protection between areas of differing intensity.

- Fourth, provide linkages between natural areas. Isolated islands of natural open space violate habitat boundaries and make the natural zones feel like captive preserves not a restoration or preservation of natural conditions.

- Fifth, the links between natural areas may be used for walking, hiking, or biking, but should be constructed of permeable and biodegradable material. In addition, the links may augment natural systems such as water flow and drainage, habitat migration patterns, or flood plain conditions.

7.3 Establish procedures that ensure the ongoing management of the natural areas as part of a strategy of sustainable development.

- Without human intervention, natural lands are completely sustainable. Cycles of growth and change including destruction by fire, wind, or flood have been occurring for millions of years. The plants and wildlife have adapted to these cycles to create a balanced ecosystem.

- Human intervention has changed the balance of the ecosystem. With the relatively recent introduction of nearby human activities, the natural cycle of an ecosystem's growth, destruction, and rebirth is not possible.

- Human settlement will not tolerate a fire that destroys thousands of acres only to liberate plant material that reblooms into another natural cycle.

■ The coexistence of human and natural ecosystems demands a different approach to design. This is the essence of sustainable design practices, a new approach that understands and reflects the needs of both natural and human communities.

CONCLUSION

We have come full circle. Ancient man understood and respected climate and built his shelters to be responsive to it. Modern man has largely ignored climate in the design of his structures. Now, once again, we realize that the land and buildings that we shape must be in harmony with nature.

LESSON 1 QUIZ

1. The south side of a building receives maximum solar radiation in the
   A. spring.
   B. summer.
   C. autumn.
   D. winter.

2. Which of the following will block the summer sun, but allow the winter sun to enter?
   I. Deciduous trees
   II. Evergreen trees
   III. North overhangs
   IV. South overhangs
   A. I and III
   B. I and IV
   C. II and III
   D. II and IV

3. When do east and west walls receive most of their solar radiation?
   A. At midday
   B. In the summer
   C. In the winter
   D. During the vernal equinox

4. As the latitude becomes more southerly, solar collectors should be
   A. closer to horizontal.
   B. tilted more.
   C. tilted more if used for heating water.
   D. oriented more westerly.

5. The amount of solar radiation on a south-facing slope is comparable to that on a
   A. north-facing slope.
   B. level site at a more northerly latitude.
   C. level site at a more southerly latitude.
   D. level site at the same latitude.

6. Which slopes tend to be cool, humid, and vegetated?
   I. Leeward
   II. North-facing
   III. East-facing
   IV. Windward
   A. I and II
   B. I and III
   C. II and IV
   D. III and IV

7. Which of the following factors does NOT affect the macroclimate?
   A. Latitude
   B. Proximity to a body of water
   C. Elevation
   D. Presence of windbreaks

8. In general, the preferred sites for residential development are
   I. at the crest of a hill.
   II. at the bottom of a hill.
   III. near a body of water.
   IV. in an area sheltered from northerly winds.
   V. in an area sheltered from southerly winds.
   A. I, III, and IV
   B. II and V
   C. III and IV
   D. I and V

9. Which of the following CANNOT be determined from a sun chart?

   A. The time the sun sets

   B. The actual number of sunny days during the year

   C. The angle that the sun makes with the horizon

   D. The position of the sun relative to true north

10. The number of degree days at a given location may be used to approximate

   A. the minimum design temperature.

   B. the amount of energy required to heat a building.

   C. the wind chill factor.

   D. the amount of insolation.

NATURAL ELEMENTS

INTRODUCTION

Each Site's Uniqueness

Just as every person has an individual identity, so too each parcel of land has its own unique character. Sites may be large or small, urban or rural—but each one is different, with its own distinctive quality and flavor.

Natural sites often evoke strong emotional responses in us. The starkness of the desert, the grandeur of the mountains, the ruggedness of the coastline—each possesses its own kind of beauty, and each creates a different mood in the observer.

Thus, there is an infinite variety of site conditions, and the unique blend of climate, topography, water, soil, rock, and vegetation that form the natural features of a site is not duplicated anywhere else. In this lesson, we will consider these natural site features, except for climate and topography, which are covered elsewhere in this course.

Manipulating the Landscape

If certain sites and site elements are beautiful and functional, while others are not, is it possible for the site planner to manipulate the landscape by removing those elements which are out of character, while enhancing those which are functional and attractive? Yes, often. New planting or other landscape elements can do wonders to heal the wounds caused by neglect or indifference. To be successful, manipulation of the landscape must always be in harmony with the essential character of the land. Does this mean that we should utilize only indigenous materials and vegetation in our site designs? Not necessarily. We simply mean that the structures and plant materials that we introduce and the landforms that we modify should have the same general character as the existing natural site elements, in order to create a pleasing and unified whole.

But some strong natural features cannot be modified. Mountains, rivers, macroclimate—these we must accept and work with, since we are unable to change them.

Visiting the Site

In discussing climate, in the previous lesson, we stressed the importance of visiting the site. Such visits are also vital if the planner is to become fully aware of its other natural features. Only by walking the site can we appreciate the views, landforms, trees, and rock outcroppings that give the site its unique character. But a site is more than the sum of its features; it has an intangible quality—a "feel"—that we try to experience during our visits. Maps and

photographs are essential, but there is no substitute for personal observation.

An effective way of recording one's impressions is to take a topographic map or survey into the field and make notes directly on it. Outstanding features, such as trees, rocks, and water bodies are particularly noteworthy, because they are important natural resources, which should be preserved if possible because of their beauty and scarcity. The aesthetic value of such features may even be the dominant reason for selecting a particular site, and the designer must fit his or her designs to them.

Objectionable elements, such as dead trees, weeds, and areas subject to earth movement or flooding are also noted. These one may remove, modify, or avoid, as may be appropriate.

Checklist

The following is an outline of natural site elements to check; of course, not all of them apply to every site.

1. Topography. See Lesson Three
2. Climate. See Lesson One
3. Water
   A. Drainage
   B. Surface water bodies
   C. Water quality
   D. Potential flooding
   E. Areas of standing water
   F. Subsurface water conditions
4. Soil
   A. Soil type and bearing capacity
   B. Depth to bedrock
   C. Depth of topsoil
   D. Areas subject to slides or subsidence
   E. Areas of erosion

5. Vegetation
    A. Species
    B. Size, form, density, and condition
    C. Color and texture
6. Views
7. Sounds and smells

EARTH AND ROCK

Introduction

Primitive man recognized four elements crucial to his survival—earth, air, fire, and water. Today, we know that there are over 100 elements, yet when we analyze the natural features of the land, we return to the primeval elements of the cave dwellers.

Geology

Geology is the science of the earth—the mountains, plains, and oceans that make up our planet. While the site planner is not expected to be a geologist, he or she should nevertheless have sufficient knowledge of the subject to understand the geological characteristics of a site. In this lesson, we will discuss the three types of rock, sedimentary, igneous, and metamorphic, and their properties.

Sedimentary Rock

Sedimentary rock, which covers most of the earth's surface, is formed by the deposition of sediments transported by streams, ocean currents, ice, or wind. When the transporting agent no longer has enough energy to carry the load, sediments are deposited and later converted to firm rock by a process called lithification. Since the sediments are laid down in beds or layers, the bedding of sedimentary rock is its most characteristic feature. Most sedimentary rock is classified as sandstone, mudstone and shale, and limestone.

Sandstone develops rugged topography because of its resistance to weathering. Shale is the most common sedimentary rock, and develops a smooth flowing topography. The surface of limestone tends to collapse and form depressions because of chemical weathering action.

Sedimentary rock is often interbedded, such as sandstone with shale or shale with limestone. The characteristics of interbedded rock vary with a number of factors, including climate and thickness of interbedding.

Igneous Rock

Igneous rock is formed when molten rock material cools and solidifies on or beneath the earth's surface. Igneous rock, such as granite, is hard, dense, and strong, and has very high bearing capacity.

Metamorphic Rock

Igneous and sedimentary rock are sometimes changed to metamorphic rock as a result of heat, pressure, and the presence of chemically active fluids—a process known as metamorphism. Metamorphic rock may be foliated or unfoliated. Foliation refers to the arrangement of the minerals in parallel layers of flat or elongated grains, along which the rock easily splits into thin flakes or slabs. Foliated metamorphic rocks include slate, schist, and gneiss, while quartzite and marble are unfoliated metamorphic rocks. Metamorphic rock occurs mainly in mountainous areas.

Weathering

The one constant in nature is its inconstancy: the earth is forever changing, albeit slowly. We are all familiar with the natural processes of weathering by which the earth's surface is being constantly worn down. When we look at the illegible inscription on an old gravestone or the crumbling face of an ancient monument, we see evidence of this ceaseless destruction. Similarly, the earth's bedrock is constantly weathered by air, water, and living matter.

Mechanical weathering, also known as disintegration, occurs as the rock is broken into smaller fragments by physical forces—expansion and contraction caused by temperature change, frost action, root pressure, and running water. Mechanical weathering, which occurs most actively in arid regions, does not alter the rock composition, only the size of the rock particles.

Chemical weathering, on the other hand, changes the rock composition by a variety of complex processes. Chemical weathering, sometimes called decomposition, is most prevalent in warm, humid areas.

Weathering is strongly influenced by climate. For example, humid climates tend to have gentle topography and deep soil, because the heavy rainfall increases weathering of the rock, while the dense vegetation acts to stabilize the weathered rock. Arid climates, conversely, tend to have rugged topography and shallow soil.

Soil

Exactly what is soil? It is a natural material, formed of decomposed and disintegrated parent rock, that supports plant life. The properties of a soil are determined by a number of factors, including the nature of the parent material, climate, topography, age, and vegetation. One type of soil will develop under an evergreen forest and another under grassland.

The profile of a soil, regardless of its type, generally consists of several horizontal layers, called *horizons.*

Below the soil is the underlying, unweathered bedrock, which may be as much as 30 feet or

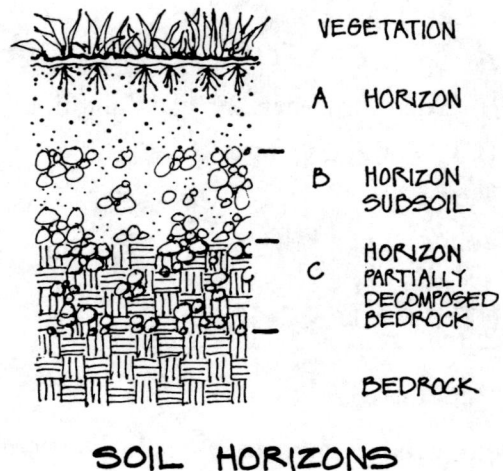

SOIL HORIZONS

more below the surface. Occasionally it surfaces as a rock outcropping. Above bedrock is the C horizon, consisting of partially decomposed rock material, which supports little, if any, plant life.

The B horizon, located directly above the C horizon, consists of material which has been further weathered and decomposed. The color varies from brownish to yellowish and some roots may be found.

The highest level is the A horizon, the soil just below the surface. It is usually dark gray in color and contains a great deal of organic matter and abundant roots. The upper six to eight inches of soil is called topsoil and contains large amounts of humus, the dark organic material produced by the decomposition of vegetable and animal matter and essential to the fertility of the soil.

The soil layers are constantly changing with time, with each horizon enlarging and encroaching into the one below it. At the same time, the surface material is slowly removed by erosion, a natural process which has been greatly accelerated by man-made developments, such as agriculture and construction.

Soil Classification

Soil means different things to different people: to architects and engineers, soil is the material that supports buildings; farmers are interested in a soil's ability to produce crops; landscape architects see soil as a plant-growing medium; while geologists study soils in order to understand the origin and evolution of the underlying rock.

Consequently, a number of different soil classification schemes have been devised. We are most interested in the classification systems based on particle size, or texture, since that largely determines three important soil qualities: drainage, bearing capacity, and erodibility. The major soil categories based on particle size are gravel, sand, silt, clay, and organic soil.

The U.S. Department of Agriculture soil classification utilizes three categories based on particle size:

1. Sand (2.0 to 0.05 millimeters diameter)
2. Silt (0.05 to 0.002 millimeters diameter)
3. Clay (smaller than 0.002 millimeters diameter)

Most soils are composed of a mixture of soil types, and have intermediate designations, such as sandy clay, silty clay, loam, etc. These designations are based on the percentages of sand, silt, and clay, by weight, contained in a sample of the soil.

Sometimes soils are given additional names, such as a geological term indicating the parent material from which the soil originated, or the name of a place where the soil type is typically found, such as Merrimac sandy loam or Fayette silt loam.

The U.S.D.A. soil classification is generally presented graphically as a triangle, in which each side is scaled from 0 to 100 percent for sand, silt, or clay, and the interior subdivided into intermediate categories, as shown in the next column.

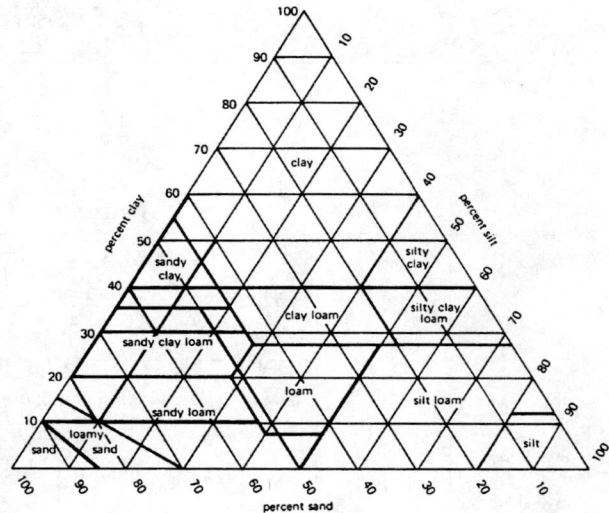

STANDARD U.S.D.A. SOIL CLASSIFICATION TRIANGLE

Although coarse-grained soils containing particles larger than sand are not included in the USDA classification, such soils often have good to excellent drainage characteristics and bearing capacity. At the other extreme and also excluded from the USDA classification are organic soils, such as peat, which have poor drainage and very low bearing capacity.

The widely-used Unified Soil Classification System groups soils into coarse-grained, fine-grained, and highly organic, using the following letter designations:

G - gravel and gravelly soils

S - sand and sandy soils

M - very fine sand and inorganic silt

C - inorganic clays

O - organic silts and clays

P_t - peat

The coarse-grained soils (G and S) are given an additional letter designation as follows:

UNIFIED SOIL CLASSIFICATION SYSTEM					
		Name	**Foundation**	**Compressibility and Expansion**	**Drainage Characteristics**
COARSE-GRAINED SOILS	Gravel and Gravelly Soils	GW well-graded gravels or gravel-sand mixtures, little or no fines	excellent	almost none	excellent
		GP poorly-graded gravels or gravel-sand mixtures, little or no fines	good to excellent	almost none	excellent
		GM silty gravels, gravel-sand-silt mixtures	good to excellent	very slight	fair to poor
		GC clayey gravels, gravel-sand-clay mixtures	good	slight	poor
	Sand and Sandy Soils	SW well-graded sands or gravelly sands, little or no fines	good	almost none	excellent
		SP poorly-graded sands or gravelly sands, little or no fines	fair to good	almost none	excellent
		SM silty sands, sand-silt mixtures	fair to good	very slight	fair to poor
		SC clayey sands, sand-clay mixtures	poor to fair	slight to medium	poor
FINE-GRAINED SOILS	Silts and Clays	ML inorganic silts and very fine sands, rock flour, silty or clayey fine sands or clayey silts with slight plasticity	fair to poor	slight to medium	fair to poor
		CL inorganic clays of low to medium elasticity, gravelly clays, sandy clays, silty clays, lean clays	fair to poor	medium	practically impervious
		OL organic silts and organic silty clays of low plasticity	poor	medium to high	poor
	Silts and Clays	MH inorganic silts, micaceous or diatomaceous fine sandy or silty soils, elastic silts	poor	high	poor to fair
		CH inorganic clays of high plasticity, fat clays	poor to very poor	high	practically impervious
		OH organic clays of medium to high plasticity, organic silts	poor to very poor	high	practically impervious
HIGHLY ORGANIC SOILS		P_t peat and other highly organic soils	not suitable	very high	fair to poor

W - well-graded (containing a mixture of particles of various sizes)

C - well-graded with clay

P - poorly-graded

The fine-grained soils (M, C, and O) have an additional letter designation as follows:

L - low compressibility and low plasticity

H - high compressibility and high plasticity

The characteristics of each soil type are shown in the table on the previous page.

Obtaining Information

How does the architect determine what soil types exist on the site? There are a number of sources of information: in agricultural areas, the Soil Conservation Service of the U.S. Department of Agriculture conducts county soil surveys, from which county soils maps are prepared. These maps show not only the soil types, but other features as well, such as water, roads, and slopes. Other sources of information include aerial photographs, geologic maps, topographic maps, and of course, personal field observation.

For more detailed soils information, particularly with regard to the soil's bearing capacity, a sub-surface exploration is necessary, as described on the following page.

Drainage

One important characteristic of a soil is its drainage, that is, its capacity to receive and transmit water. As one can see from the table on page 26, coarse-grained soils (sands and gravels) generally have good drainage, while fine-grained soils (silts and clays) often drain poorly.

Vegetation improves the drainage of a soil. Thus, a vegetated sandy soil absorbs most of the rain that falls on it and consequently, surface runoff is low. At the other extreme is a barren clayey soil, which drains so poorly as to be almost impervious, causing heavy surface runoff.

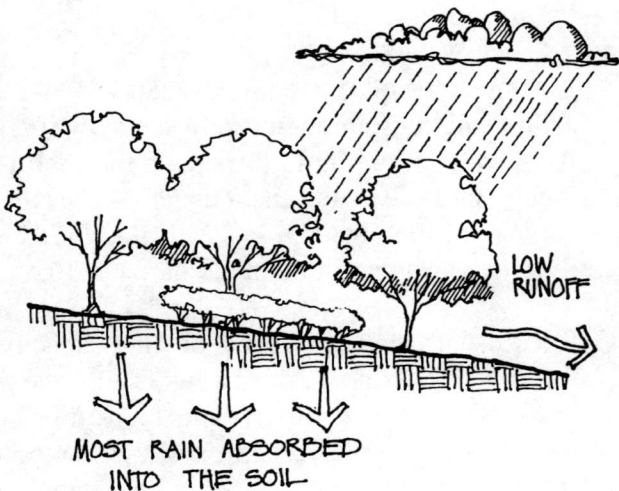

LOW RUNOFF

MOST RAIN ABSORBED INTO THE SOIL

VEGETATED SANDY SOIL

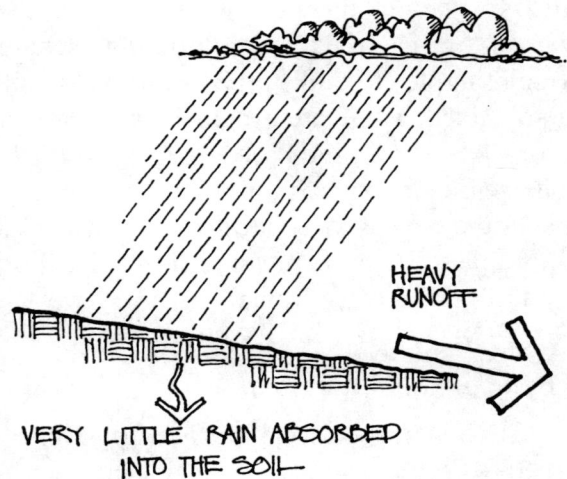

HEAVY RUNOFF

VERY LITTLE RAIN ABSORBED INTO THE SOIL

BARREN CLAYEY SOIL

VARYING DRAINAGE CHARACTERISTICS

Poorly-drained soils may cause flooding, since they cannot absorb much rain water. Paving also increases the possibility of flooding, because it prevents water from infiltrating into the ground. Where we have a choice, it is preferable to pave over impermeable soil, which cannot absorb much rain water anyway, rather than over well-drained soil.

Bearing Capacity

The bearing capacity of a soil may be determined by several different methods. If one knows the soil type, the values given in the building code may be used. A typical example is Table 18-I-A of the Uniform Building Code, which is reproduced above right by permission of the International Conference of Building Officials.

The use of such tables is usually limited to relatively small, lightly-loaded structures. For other buildings, a subsurface exploration is often necessary. For example, test pits may be excavated, to expose the subsurface soils for in-place examination.

The most common subsurface exploration consists of a series of exploratory borings, from which samples of undisturbed subsurface soils are obtained. These are examined and tested in a soils laboratory, and based on these tests, the soils engineer prepares a report which recommends the type of foundation to be used and the allowable soil bearing pressure. Two criteria are used to establish this value:

1. Shear failure of the soil, as shown in the next column must not occur.

2. The building must not settle excessively. Some settlement always takes place and is unavoidable; however, the total settlement of the building, and the differential settlement between portions of the building, should be limited to some small amount which the structure can tolerate.

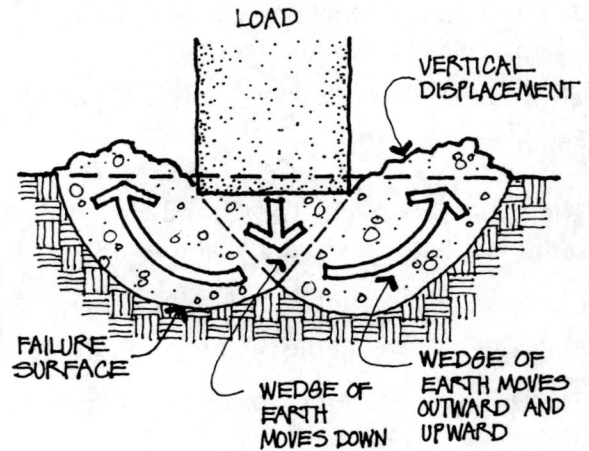

SOIL FAILURE

How many borings should be made? That depends on many factors, but at least four borings close to the corners of the building are required, with additional borings about 50 to 100 feet apart.

How deep should the borings be? Again, there are no exact rules, but at least deep enough to penetrate soft or unsuitable material and extend into soil of adequate strength.

A detailed log of each boring is made, and these soil boring logs are included in the soil report. A typical soil boring log is shown on the following page. The designations used (SM, SP, etc.) conform to the Unified Soil Classification System, as shown in the table on page 30.

Rock usually has the highest bearing capacity, followed by coarse-grained soils (sands and gravels). Clays and silts are fair foundation materials, and finally, organic soils, soft or loose soils, or unconsolidated fills are considered unsuitable to support buildings.

Sites underlain by soft or loose soils are well-suited to open space uses, such as recreation,

| | | **Lateral Bearing Lbs./Sq./Ft./ Ft. of Depth Below Natural Grade[4]** | **Lateral Sliding[1]** | |
Class of Materials[2]	**Allowable Foundation Pressure Lbs./Sq. Ft.[3]**		**Coefficient[5]**	**Resistance Lbs./Sq. Ft.[6]**
1. Massive Crystalline Bedrock	4000	1200	.70	
2. Sedimentary and Foliated Rock	2000	400	.35	
3. Sandy Gravel and/or Gravel (GW and GP)	2000	200	.35	
4. Sand, Silty Sand, Clayey Sand, Silty Gravel and Clayey Gravel (SW, SP, SM, SC, GM and GC)	1500	150	.25	
5. Clay, Sandy Clay, Silty Clay and Clayey Silt (CL, ML, MH and CH)	1000[7]	100		130

ALLOWABLE FOUNDATION AND LATERAL PRESSURE

[1] Lateral bearing and lateral sliding resistance may be combined.

[2] For soil classifications OL, OH and PT (i.e., organic clays and peat), a foundation investigation shall be required.

[3] All values of allowable foundation pressure are for footings having a minimum width of 12 inches and a minimum depth of 12 inches into natural grade. Except as in Footnote No. 7 below, increase of 20 percent allowed for each additional foot of width or depth to a maximum value of three times the designated value.

[4] May be increased the amount of the designated value for each additional foot of depth to a maximum of 15 times the designated value. Isolated poles for uses such as flagpoles or signs and poles used to support buildings which are not adversely affected by a 1/2-inch motion at ground surface due to short-term lateral loads may be designed using lateral bearing values equal to two times the tabulated values.

[5] Coefficient to be multiplied by the dead load.

[6] Lateral sliding resistance value to be multiplied by the contact area. In no case shall the lateral sliding resistance exceed one half the dead load.

[7] No increase for width is allowed.

agriculture, or parking. However, building foundations may not be supported on such soils, but must extend through the soft material and into soil or rock of adequate bearing capacity. This is usually done in one of two ways: by driving piles of wood, steel, or concrete; or by drilling holes and filling them with concrete. Another approach which is sometimes used is to remove the unsuitable soil and replace it with properly compacted fill. The soil used for such fill may be the on-site soil, or it may be soil brought from another location and compacted in place. The compaction is done mechanically, often by sheepsfoot rollers, and must be closely inspected and tested to be sure that the proper degree of compaction is obtained.

Shrinkage and Swelling

In most areas of the country, moisture in the soil freezes in the winter, causing the soil to expand. This is harmless if the site is undeveloped. But foundations in such soil are likely to heave and cause structural damage. To minimize this possibility, building footings should be placed below the frost line, which is the deepest penetration of frost expected. The depth of the frost line varies from zero to as much as 100 inches, as shown in the map on page 36.

Expansive soils are found in many parts of the country. These are clays which swell when they become wet and shrink when dried. Such volumetric changes can cause extensive damage to building footings and slabs on grade. Footings in such soils should be extended below the depth of seasonal moisture change, so that the moisture content of the subsoil remains fairly constant: it is the *change* of moisture content that causes the problem.

Sometimes, instead of extending continuous footings to the required depth, spaced foundation piers are used, which must be insulated from the surrounding expansive soil.

Erosion

Erosion is the process by which the surface of the earth is worn away by the action of natural elements, such as water and wind. A site is particularly vulnerable to erosion during construction, when the land is usually stripped of vegetation and the soil is exposed to rain and wind.

Although erosion cannot be eliminated altogether, it can be controlled. Some of the measures which can be taken are as follows:

1. Disturb as little area as possible.
2. Do not remove any planting unless absolutely necessary.
3. Stockpile and protect topsoil. This valuable soil has taken thousands of years to produce and should be retained for use after construction.
4. Provide temporary dams and channels to slow down runoff and collect eroded soil.
5. Leave soil exposed for as short a time as possible.
6. Avoid steep banks.
7. Replant exposed areas as soon as feasible.

Erosion is not limited to sites undergoing construction, however; natural and developed sites may also be susceptible to erosion, depending on four factors:

1. The type of soil
2. The steepness of the slopes
3. The vegetative cover
4. The speed and volume of runoff

Sandy soils tend to be eroded more than clayey or gravelly soils—that is, they have less resistance to the erosive power of flowing water. Steep slopes erode faster than shallow slopes, and sites with little or no vegetation are more susceptible to erosion than those that are heavily planted. Where the velocity and volume of storm water runoff are great, erosion is more likely to occur. Thus, the steepest part of a slope usually experiences the greatest erosion and has only a thin layer of topsoil. The material eroded is deposited near the bottom of the slope, resulting in a thick layer of topsoil.

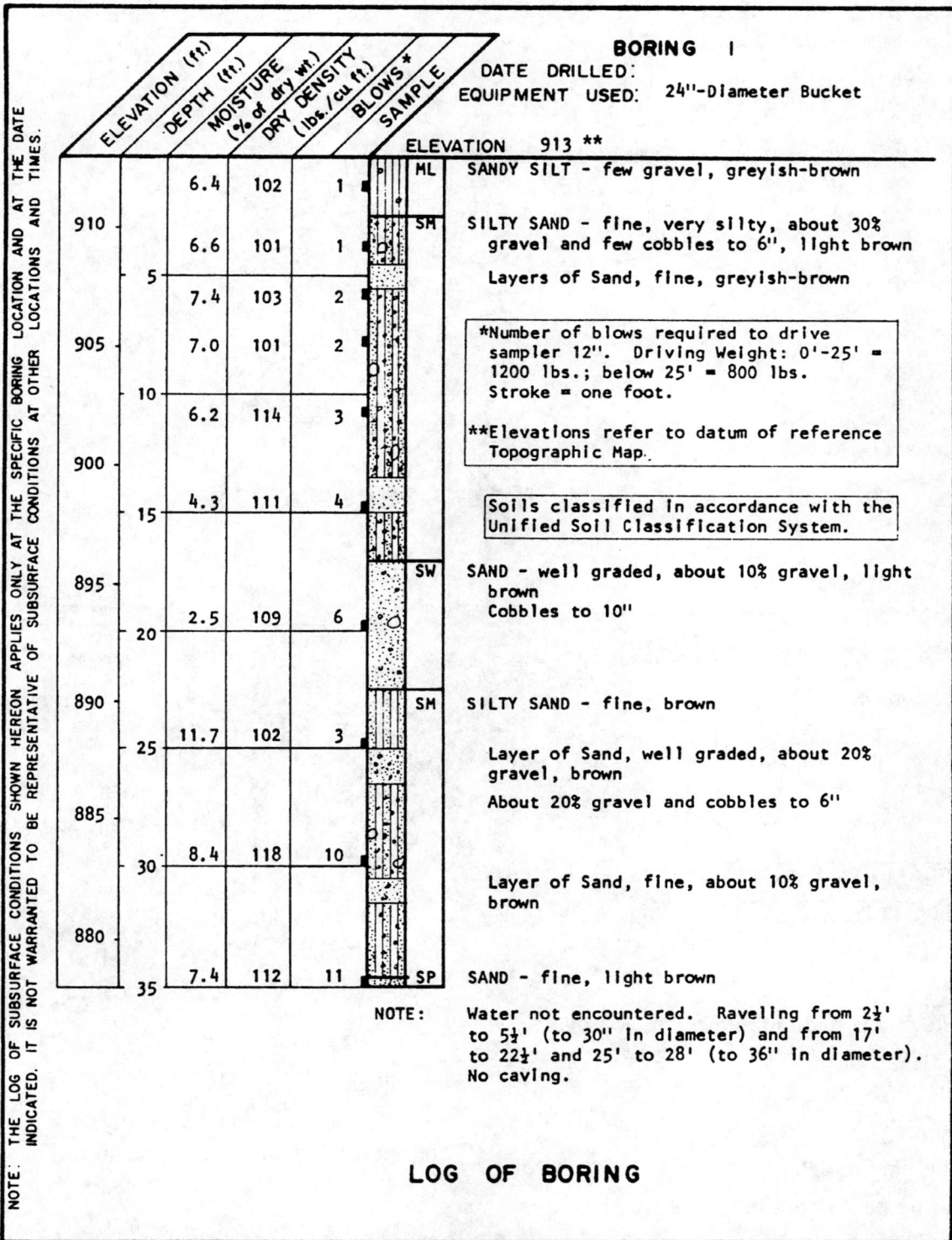

BORING 1

DATE DRILLED:
EQUIPMENT USED: 24"-Diameter Bucket

ELEVATION 913 **

ELEVATION (ft.)	DEPTH (ft.)	MOISTURE (% of dry wt.)	DRY DENSITY (lbs./cu ft.)	BLOWS *	SAMPLE	Description
		6.4	102	1	ML	SANDY SILT - few gravel, greyish-brown
910		6.6	101	1	SM	SILTY SAND - fine, very silty, about 30% gravel and few cobbles to 6", light brown
	5	7.4	103	2		Layers of Sand, fine, greyish-brown
905		7.0	101	2		
	10	6.2	114	3		
900						
	15	4.3	111	4		
895					SW	SAND - well graded, about 10% gravel, light brown. Cobbles to 10"
	20	2.5	109	6		
890					SM	SILTY SAND - fine, brown
	25	11.7	102	3		Layer of Sand, well graded, about 20% gravel, brown. About 20% gravel and cobbles to 6"
885						
	30	8.4	118	10		Layer of Sand, fine, about 10% gravel, brown
880						
	35	7.4	112	11	SP	SAND - fine, light brown

*Number of blows required to drive sampler 12". Driving Weight: 0'-25' = 1200 lbs.; below 25' = 800 lbs. Stroke = one foot.

**Elevations refer to datum of reference Topographic Map.

Soils classified in accordance with the Unified Soil Classification System.

NOTE:　Water not encountered. Raveling from 2½' to 5½' (to 30" in diameter) and from 17' to 22½' and 25' to 28' (to 36" in diameter). No caving.

LOG OF BORING

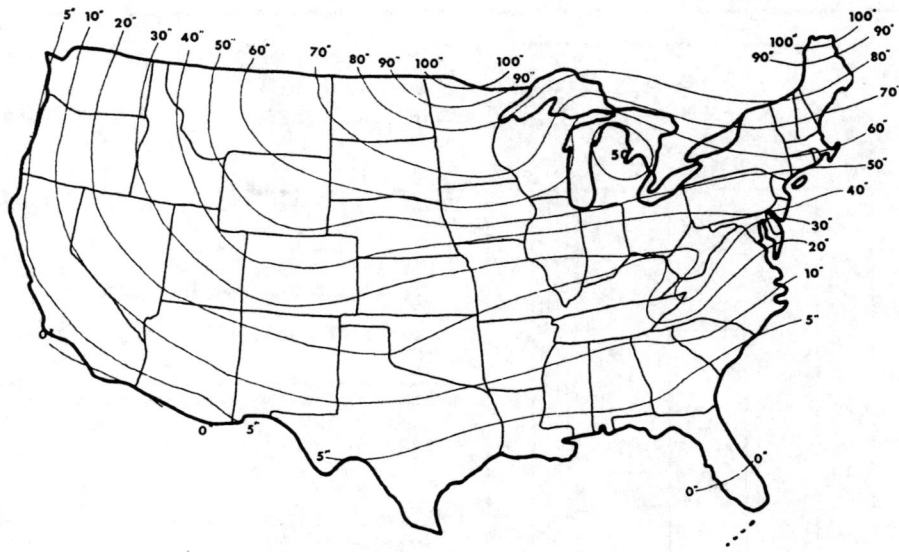

MAXIMUM FROST PENETRATION

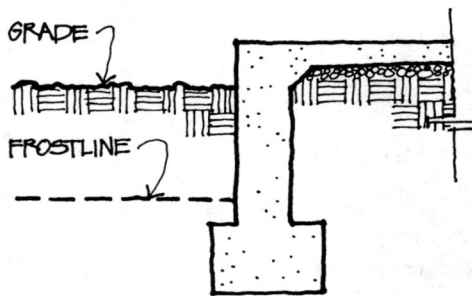

FOOTING EXTENDS BELOW FROSTLINE

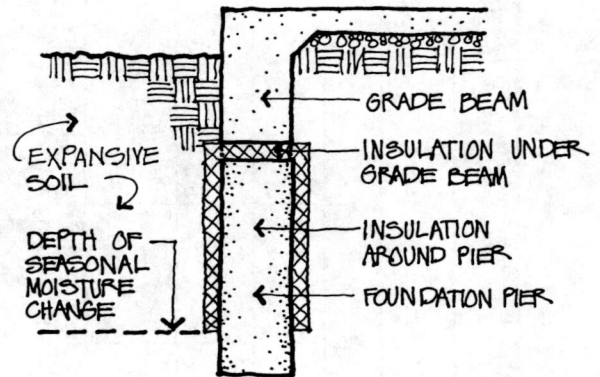

FOUNDATION PIER IN EXPANSIVE SOIL

The site planner can minimize erosion by employing a variety of slope stabilization techniques, including:

1. Planting
2. Applying mulch
3. Facing banks with rubble or riprap
4. Retaining banks with cribbing or retaining walls

Sliding

Earth slides can occur naturally, or as a result of improper grading. The potential slide hazard is a function of several factors:

1. The steeper the slope, the more likely it is to slide.

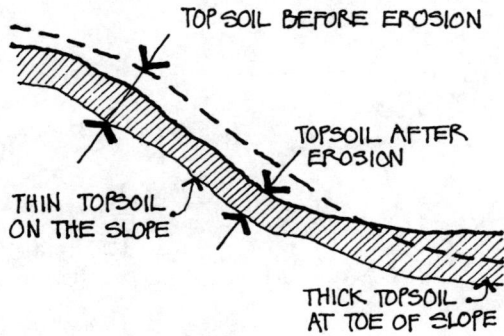

SLOPE EROSION

2. Fine-grained soils (clays and silts) are more susceptible to sliding than coarse-grained soils (sands and gravels), and firm bedrock is the least likely to slide.

3. The flow of water may trigger a slide, because it increases the weight of the soil and lubricates the plane between soil types.

4. Layered soils slide more readily than homogeneous soils.

5. Undercutting an existing slope, whether by natural processes or for construction, may increase the possibility of sliding.

How can we minimize the possibility of earth slides? One obvious step is to consult a soils engineer before developing any area where slides might occur, that is, any area containing slopes whose underlying soil structure is unknown. Areas known to be prone to sliding should, of course, be avoided for building development. Since water often provides the triggering

1. PLANTING

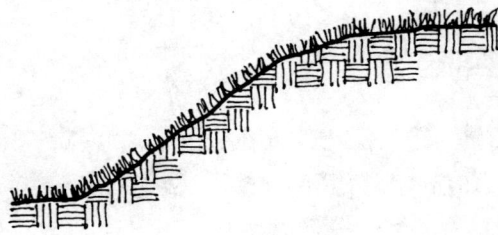

2. MULCH

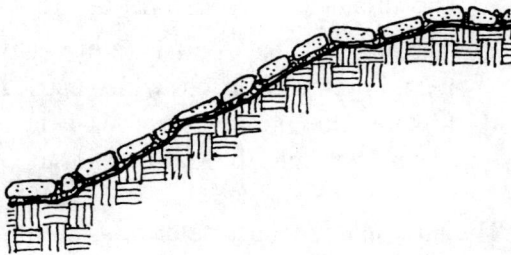

3. RUBBLE or RIPRAP

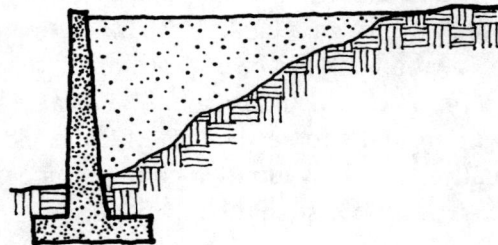

4. RETAINING WALL

SLOPE STABILIZATION METHODS

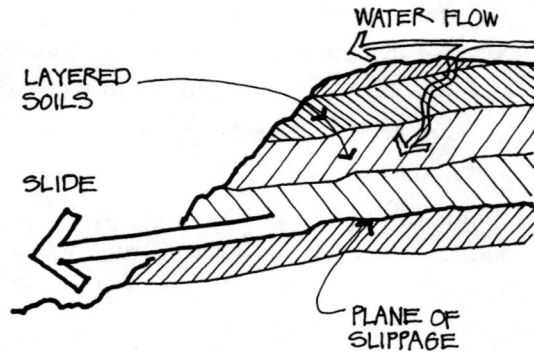

EARTH SLIDE CAUSED BY WATER FLOW

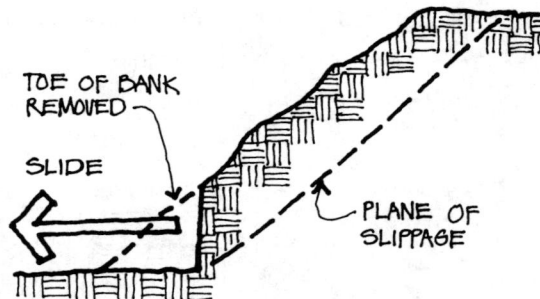

EARTH SLIDE CAUSED BY UNDERCUTTING OF BANK

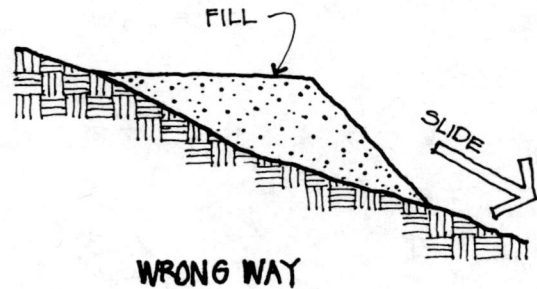

WRONG WAY

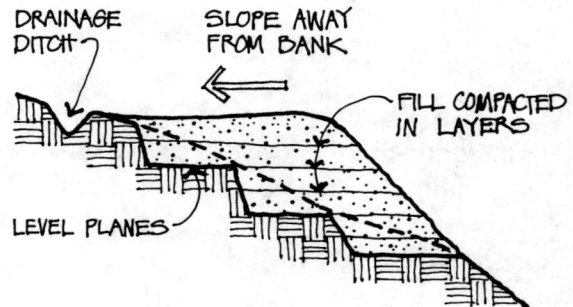

RIGHT WAY

IMPROPERLY PLACED FILLS CAN CAUSE SLIDING

mechanism for a slide, it may be advisable to intercept the flow of storm water by a drainage device of some kind, so that water is not allowed to flow on or through a bank. Even where construction near a bank has been determined to be safe and feasible, structures should be located well away from the top or bottom of the bank. And finally, any excavation which might undermine existing slopes should be avoided.

Fills placed on a slope tend to slide. To reduce this possibility, fills should be placed on level planes, as shown below.

Subsidence

In some locations, subsidence, or sinking of land, has occurred. There are two basic causes of subsidence:

1. The presence of highly compressible subsoils, such as loose fill or organic soil.
2. The pumping out of oil, natural gas, or water, resulting in void spaces which cause the land to sink. One possible solution is to pump water back into the soil to fill the voids.

The site planner must determine if subsidence is a factor in the site under consideration. While subsiding areas are satisfactory as sites for recreation or other open space uses, they should be avoided as sites for structures.

Fertility

Site planners are frequently concerned with the ability of a soil to support vegetation. This, in turn, depends on four major characteristics of the topsoil: drainage, humus content, acidity, and the presence of nutrients. These characteristics can be determined by tests, and deficiencies can frequently be remedied, so that most soils can be made suitable for plant growth.

Fertilizers can be added to soil with insufficient humus, impervious soils can be broken up to improve drainage, acidity can be modified, and nutrients (potassium, phosphorus, and nitrogen) can be added to soils deficient in them. If the existing soil is particularly poor as a plant-growing medium, topsoil may be brought onto the site and spread over the ground.

WATER

Introduction

Water is essential to all life. Yet in most parts of the United States and other highly developed countries, it is taken for granted, like the air we breathe. We turn on the faucet and it gushes forth; we clean, cook, bathe, flush, and irrigate with it, with little or no thought given to its source or continued availability.

But water is more than a physical necessity—it is a vital part of the landscape both aesthetically and emotionally. Since time immemorial, water has had a tremendous appeal for people. Whatever its form—pool, river, fountain, or waterfall—water is one of the most fascinating of all natural design elements.

Uses of Water

Water has many functions in site design. Some are aesthetic in nature—the still water of a lake

is soothing and evokes a feeling of serenity. However, the water body need not be natural to have a strong impact; the rigid geometry of a reflecting pool may also provide a contemplative setting. In contrast to the tranquility of still water, the swift moving water of a fountain or waterfall is dramatic and exciting, both visually and aurally.

Water has a wide variety of practical uses as well. Like all living things, plants and lawns need water to survive and flourish, and in many areas of the country, this is provided by sprinkler systems, which may be designed or specified by the site planner. Water also moderates the microclimate of a site, as discussed in the previous lesson.

The sound of falling water may be used to mask urban noise from cars and other sources, as in Freeway Park in Seattle.

The recreational uses of water are many and varied: swimming in a backyard pool, sailing or waterskiing on a lake, snorkeling in the ocean. The site planner may have to consider the conflicting needs of recreational users, i.e., swimming vs. power boating.

The need for water as recreation may have to be weighed against environmental and other considerations: should a wild river be tamed by damming, or should it be left alone, for scenic enjoyment and white water rafting? Some of these decisions are made by the site planner, while others are beyond his or her control. But in any event, the planner should have a thorough understanding of water as a natural resource.

Water in Site Design

Wherever a body of water exists, the land near it is very desirable, and this is reflected in the high price of waterfront property. It seems reasonable, therefore, that any body of water on a site should be preserved, protected, and enhanced. Let's examine each of these goals.

One should preserve a water body by leaving it and the area surrounding it in the natural, undisturbed state whenever possible. We protect it by preventing any kind of contamination. For example, polluted surface runoff should be treated or filtered before being allowed to flow into a body of water. We also maintain natural drainage channels whenever possible and provide detention swales or ponds to prevent flooding.

How can we enhance a body of water? One way is to limit development along the shoreline, thereby creating attractive open space, as well as a much longer effective shoreline set back from the water.

Enhancement of an existing lake may also come about when a dam is built. Dams are constructed for a variety of purposes, including flood control and generation of hydroelectric power. One

by-product is the creation or enlargement of an upstream lake, which can be used for recreation and as a scenic feature. In recent years, many people have questioned the wisdom of large-scale dam construction, because it inevitably affects, and may even destroy, the natural ecology of an area. Here, as in many other aspects of site and regional planning, one must weigh the advantages and disadvantages of man's intervention in nature.

Just as buildings can become dilapidated through age, use, or misuse, so too some water bodies become unattractive and unappealing: the canals in Venice, California, which were once intended to rival those of their Italian namesake, have fallen into disrepair through neglect. But such water features can be reclaimed and restored.

Where a water body is created or reclaimed, it is often desirable that its shape be natural and curvilinear, rather than artificial and geometric. Of course, this is not always the case. Where man-made forms predominate, as in many urban settings, it is often appropriate for a body of water introduced into the environment to be rigid in shape and appear man-made.

If possible, the shape of a lake or pond in a rural setting should be such that the entire pond cannot be seen from the shoreline, thus adding to the sense of mystery and the appeal of the pond. Usually, the best sense of balance and harmony is attained when the lake or pond is lower in elevation than any other point in the immediate area.

Paths along the shore should reflect the shape of the pond or lake and the undulation of the water. The design of paths, bridges, docks, and any other structures at the water's edge should be simple and utilize durable and water-resistant

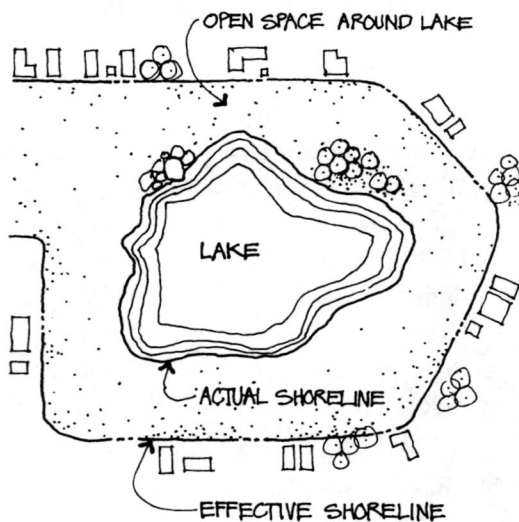

LIMITING SHORELINE DEVELOPMENT

RIGID GEOMETRIC SHAPE

SURROUNDING WALLS
AND BUILDINGS

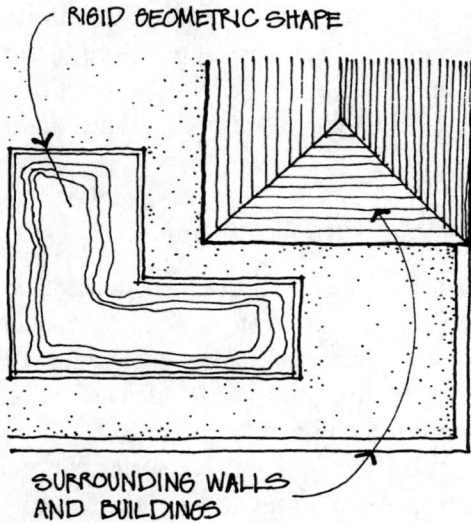

REFLECTION POOL IN URBAN SETTING

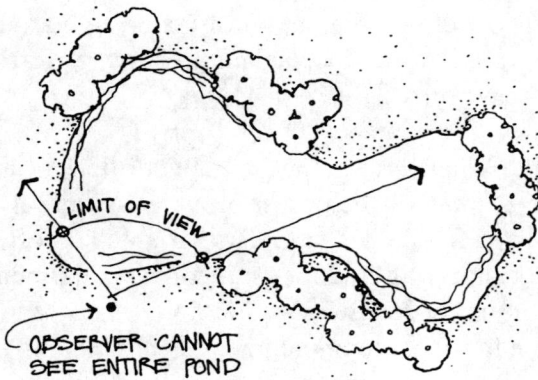

LIMIT OF VIEW

OBSERVER CANNOT
SEE ENTIRE POND

POND IN RURAL SETTING

materials. Corrosion and weathering are always problems near the water, particularly in the salty atmosphere close to the ocean.

Where possible, banks should be left natural, unless they require surface treatment to withstand the erosive effect of the water.

Such surface treatment may consist of stone, reinforced concrete, treated lumber, or steel,

always allowing sufficient freeboard (distance from normal water line to the top of the adjacent surface) for the highest expected water level and maximum wave action.

Whatever the size of the site, the use of water as an outdoor design element adds interest and symbolizes refreshment.

Streams

A stream is any body of water flowing in a channel, such as a river or brook. Its flow varies with the year, the season, and the place, but a stream is always part of a natural drainage system, and therefore, it should be disturbed as little as possible. In general, river banks should also

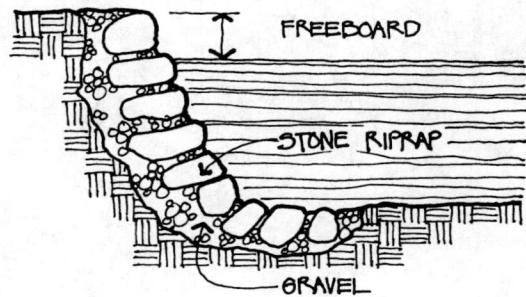

FREEBOARD

STONE RIPRAP

GRAVEL

STONE

GRAVEL
BACKFILL

REINFORCED
CONC. WALL

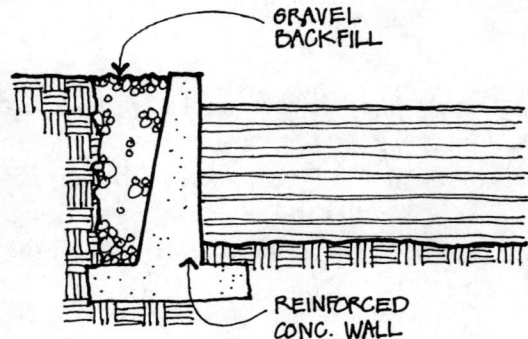

REINFORCED CONCRETE

REINFORCING THE WATER'S EDGE

be left alone, because reshaping them or removing existing vegetation may increase erosion.

But nature does not always serve man's needs, and sometimes alterations are necessary. For example, rivers must be crossed. Such crossings should be located where they are most feasible structurally: where the stream is narrow, to minimize the length of span; where the banks are stable, so that economical foundations can be constructed; and where the banks are higher than the highest expected flood line.

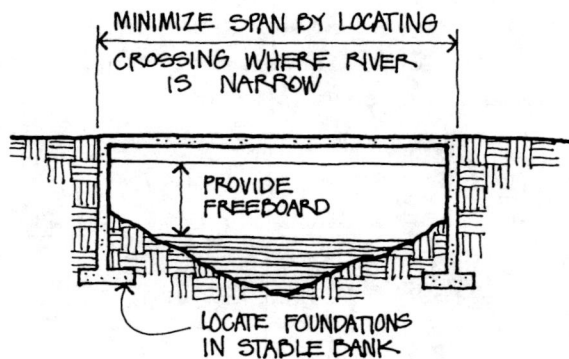

RIVER CROSSING

If a crossing must be made in an area subject to flooding, the bridge members should be designed to resist the dynamic action of flood waters.

Where the span between banks is great, additional piers within the stream may be required. Such piers should be oriented with their long dimension parallel to the direction of flow, to minimize disruption of stream flow and the resulting turbulence.

An open man-made drainage channel is also a stream, usually lined with concrete. Such a channel is most efficient if it is straight, without curves or bends, and with a constant width and depth. But a straight channel with a uniform cross-section is not very interesting; a curvilinear channel with a varying cross-section and landscaped banks has a more natural and attractive appearance.

Waterfalls and Fountains

Water falling freely through space because of a sudden change in elevation of its channel creates the most dramatic of all water displays—the waterfall. Natural and man-made waterfalls occur in the outdoor environment in a variety of sizes, shapes, and descriptions. The water may fall in a smooth sheet, or it may be rippled. There may be several falls in segments, or one free fall. And the water may fall into a lake or pool at its base, or onto a hard surface, such as rock.

Each waterfall is unique and creates a different effect. But the interaction of water, light, and sound is always spectacular and always forms a focal point.

The fountain is another dramatic, often theatrical, display of the power of water. Where it is utilized in site design, a large fountain is often the center of attraction, sometimes comprising a variety of jets with multicolored lights and even musical accompaniment. But fountains are not always spectacular—a small, simple fountain, for example, can provide a point of interest in a backyard garden.

Unlike waterfalls, fountains are almost always man-made, with the exception of natural geysers. Regardless of its size or shape, a fountain is perceived as a cool element, making it particularly attractive in a warm, dry climate.

Water Cycle

All the water on the earth, under the ground, and in the atmosphere is part of one unified

PRECIPITATION

EVAPORATION   TRANSPIRATION   EVAPORATION

RUNOFF

OCEAN

INFILTRATION

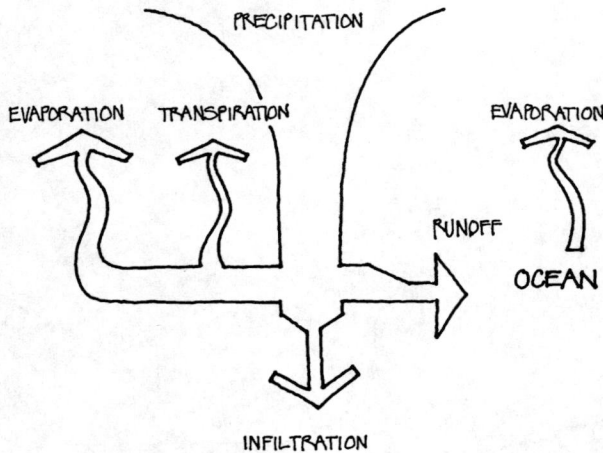

DIAGRAMMATIC WATER CYCLE

system, called the hydrologic or water cycle. After water falls on the land as *precipitation*, either rain or snow, it follows three paths. A small amount flows off the land into streams and eventually into the ocean as *runoff*. A still smaller quantity soaks into the ground as *infiltration*. Most of the precipitation is evaporated into the atmosphere directly and by *transpiration*, through the action of plants, thus completing the cycle.

Runoff therefore consists of total precipitation, less the water infiltrated into the soil, less the water evaporated directly to the air, less the water transpired back to the air from plants, as shown in the diagram above.

In site planning, we are concerned with all of these hydrologic processes, particularly runoff and infiltration. When a site is developed, the amount of runoff increases. To understand why this is true, let's look at the water cycle diagram. Site development generally entails the removal of some vegetation, thus decreasing the amount of transpiration. Also, relatively pervious land is replaced by impervious buildings, streets, and parking areas, which reduces infiltration. Less

transpiration into the air and less infiltration into the soil mean more surface runoff.

How does the site drainage system handle this surface runoff? One approach is for the runoff to immediately enter the drainage system. A different approach requires that most rainfall be held in a *detention pond* on the site until the rain subsides. The runoff is then released slowly without causing flooding. The intent of this second approach, which is often required by local ordinance, is that the flow of rain water from the new development be equal to the runoff from the site prior to development.

Potential Flooding

We have said that the development of a site increases the amount of runoff. On a larger scale, urbanization has a similar effect on the hydrologic cycle: the amount and speed of runoff are increased, the runoff is warmer and contains pollutants, and the stream which eventually carries the runoff is visually impaired as a result of erosion and pollution. Urbanization in this country has been especially rapid during the past 30 years. Consequently, there have been a number of disastrous floods. Although measures have been taken by various public agencies, the potential danger of flooding remains high in some areas.

The relatively flat land within which a stream flows is called a *flood plain*. When the volume of flow exceeds the stream's capacity, it overflows its banks and spreads onto the flood plain. This occurs regularly, and the boundary of the flooded area depends on the flood frequency. Thus a ten-year flood inundates less land than a 100-year flood.

The term ten-year flood refers to a flood of a magnitude such that it is likely to occur only once every ten years; therefore, the likelihood

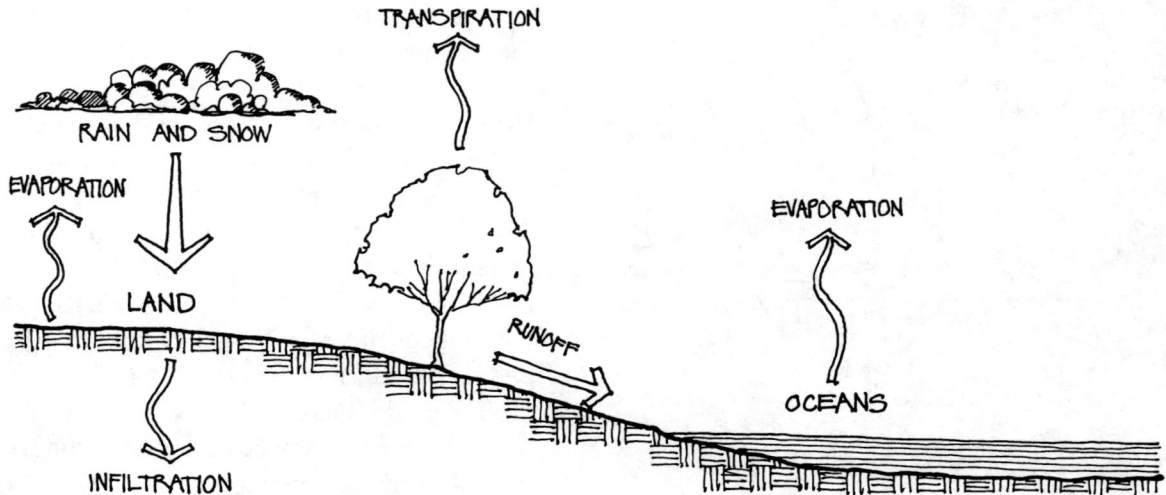

THE WATER CYCLE

of a ten-year flood occurring in any given year is ten percent. Similarly, a 100-year flood is one that is expected to occur once every 100 years. A 100-year flood therefore has a much greater magnitude than a ten-year flood. When designing in flood-prone areas, we select a flood of a given magnitude—say a 100-year flood—and set floor elevations above that flood level.

Since the flood plain is subject to natural and recurring floods, it seems obvious that any construction within the flood plain courts disaster.

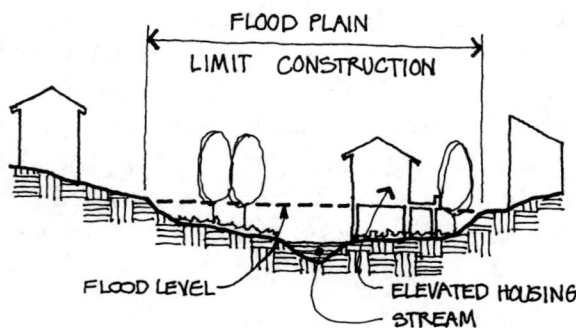

SECTION THROUGH FLOOD PLAIN

Therefore, such lands should be limited to open space uses, such as recreation and agriculture. This swath of land can provide a natural, park-like, easily-maintained setting. Unfortunately, it doesn't always work out that way; as desirable land for development becomes scarcer and more expensive, there is increased pressure to build on flood plains. As a compromise, low-density housing is often permitted, provided the occupants are aware of the potential hazard and the structures are elevated above flood level.

The water table in a flood plain usually occurs near the surface, drainage is generally poor, and the soil deep and uniform. The soil is often subject to large volumetric changes when it becomes wet, making it unsatisfactory for supporting building loads, but usually excellent for agriculture. The rivers in flood plains are often meandering.

The conventional solution to the problem of potential flooding involves the construction of concrete channels. Alternatively, existing natural drainage channels and flood plains can be utilized, even if some modifications are required. Such solutions preserve the aesthetics and ecology of the natural environment.

Underground Water

The water contained in the voids and crevices under the earth's surface exceeds by far all of the water contained in streams and lakes. This underground water comes from precipitation, both rain and snow, which soaks directly into the ground or drains into rivers and lakes and then seeps into the ground.

Underground water is found either in the zone of aeration or the zone of saturation. The zone of aeration is the higher zone, where the spaces between the soil grains contain both water and air. In the lower zone, the zone of saturation, all of the void spaces are filled completely with ground water. The irregular surface which forms the boundary between the two zones is called the *ground water table*. This is usually a sloping surface which fluctuates seasonally and roughly follows the ground surface.

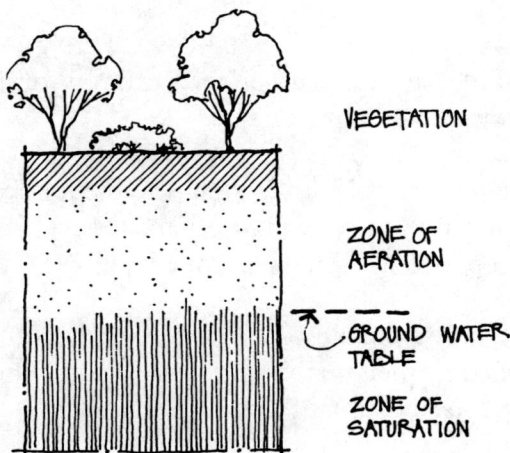

VEGETATION

ZONE OF AERATION

GROUND WATER TABLE

ZONE OF SATURATION

UNDERGROUND WATER ZONES

Where the ground water table is high, about six feet below the surface, construction excavation must be braced and kept dry by pumping. Basements must be waterproofed, basement walls designed to resist hydrostatic pressure, underground tanks or other structures designed to resist uplift, and the bearing capacity of the foundation soils often reduced.

Underground water flows at a very slow velocity, depending on the porosity and permeability of the underground earth or rock material. An underground permeable material through which water flows is called an *aquifer*. Sand, gravel, sandstone, and some limestones are generally good aquifers, while clay, shale, and most metamorphic and igneous rocks are poor aquifers.

PLANTS

Introduction

Plants are an important site design element and provide beauty and vitality to the outdoor environment. A site development without plants would be like a moonscape—stark and lifeless. Plants soften the hard edges, define spaces, and add interest.

The use of planting in site design is not merely ornamental or decorative, any more than a building design is. Rather, the site planner often considers planting as a functional element and an integral part of the overall design of the site.

Plants are unique in that they are the only design element that is alive and hence ever-changing. They grow, change with the seasons, and move with the winds. The site planner must be aware of seasonal characteristics and growth patterns, and realize that the plant that is placed in the ground today will look different next month and next year.

Other design elements can often be ignored after a site is developed, but plants, being alive, need constant nurturing; they need the right soil, sun and wind exposure, temperature range, moisture, and nutrients to live and thrive. The site planner considers this need for ongoing maintenance and often selects plant materials that are relatively easy to care for. On the other hand, where proper maintenance of vegetation is simply too difficult or expensive to achieve, other elements, sometimes inferior, are used as substitutes—for example, blacktop paving instead of grass in a school yard.

Like all natural things, plants are imperfect and not totally predictable. That is the essence of their appeal: they provide a connection with nature.

Defining Space

When we speak of plants or vegetation in this lesson, we are referring to all living organisms in the environment which draw their sustenance from the soil. Trees, shrubs, ground cover, lawn—all are plants. Our discussion comprises both indigenous vegetation and plants introduced into the environment. Since native plants, by their very existence, are well suited to a site, they should be preserved, unless there is an overriding reason to remove them.

New plants should not be introduced haphazardly, but only after careful consideration. A well thought out landscape can change an ordinary site into one that is attractive and distinctive.

In addition to their aesthetic value, plants serve a variety of other functions in the outdoor environment, including defining space. In a building, space is usually defined by rigid physical elements, such as walls, ceiling, and floor. Outside, the definition of space is more subtle. Trees or shrubs may provide a feeling of vertical

CEILING CREATED BY TREES

TREES CREATE HORIZONTAL ENCLOSURE

enclosure, without actually enclosing an area. With deciduous trees, the spatial definition is much stronger in the summer than in the winter, when the trees have lost their foliage. Closely-spaced trees may also provide a horizontal enclosure, or ceiling.

In addition to forming enclosures, trees may visually connect structural elements, such as buildings, and direct people into a space. Thus, the site planner can create a variety of spatial feelings through the use of plants.

Plants can act as a screening device; a cluster of tall, dense trees may provide privacy for an outdoor terrace, for example.

While the trees block the view into the terrace, they also prevent viewing from the terrace, and thus separate it from its surroundings.

Unattractive site elements may be visually screened by planting: most people would rather look at trees than mechanical equipment or parked cars.

Of course, planting used for privacy or screening should be evergreen, in order to be effective throughout the year.

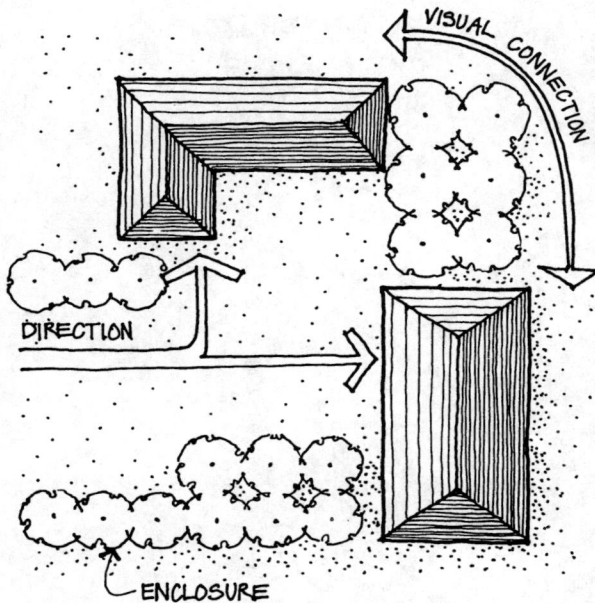

TREES ENCLOSE, DIRECT, AND CONNECT

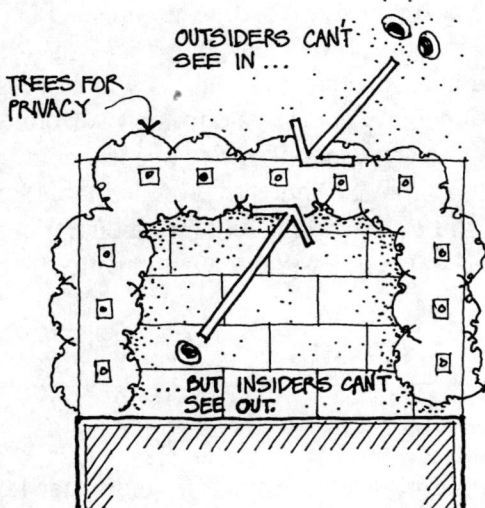

PRIVACY CREATES SEPARATION

Environmental Control

Vegetation is one of the most moderating influences on the environment. Trees block both the sun and the wind. They act as nature's air conditioning by cooling, humidifying, and filtering the air. They create sheltered zones by reducing wind speeds.

Trees and other planting help to control erosion, destructive runoff, and flooding. They absorb sound. And they provide a habitat for birds.

In these and other ways, plants improve the quality of the environment and hence, the quality of human life.

Aesthetics

For convenience, we have separated the various functions of plants. In reality, of course, all of these functions may be performed simultaneously; a tree can provide shade, help to define space, and look beautiful, all at the same time.

Trees are the dominant plant material; they must be carefully selected and placed to assure a successful design. In general, trees should be clustered as they are in nature, and not spaced too regularly or too far apart. A row of uniformly spaced trees tends to look formal and unnatural; however, it may be appropriate along a street or in an urban setting, to reflect the rigid forms of the built environment.

Occasionally, a single tree can be used as a focal point in the outdoor environment, in much the same way as a piece of sculpture. It may stand by itself, or it may be complemented by smaller trees and plants to create a unified composition.

While the larger trees are dominant, smaller trees and shrubs are used to subdivide the site into smaller areas, visually connect the various site elements, define paths or roads, and add visual interest.

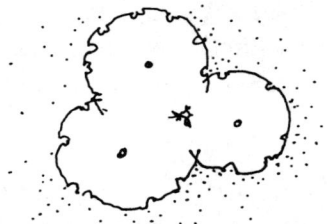

INFORMAL & NATURAL

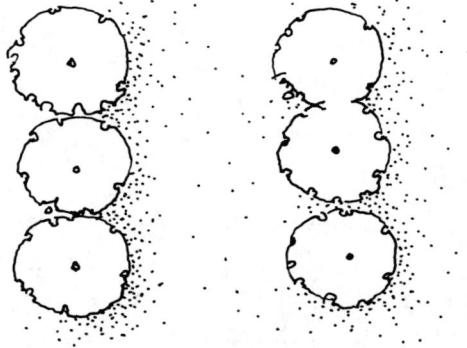

FORMAL & UNNATURAL

TREE GROUPINGS

If we think of larger plant materials forming the walls and ceiling of the outside environment, then ground cover is its carpet. Ground cover defines a space or surface, provides visual interest because of its color or texture, and retains soil and moisture.

Trees or other plant materials may be used to frame a view.

Just as the facade of a building should be free from clutter, without too many materials or fussy details, so too one should avoid the use of numerous plant varieties and complex groupings. It's best to keep it simple, using just a few carefully chosen, grouped, and placed plant varieties.

However, this does not mean that only one plant size or form should be used. There is an almost

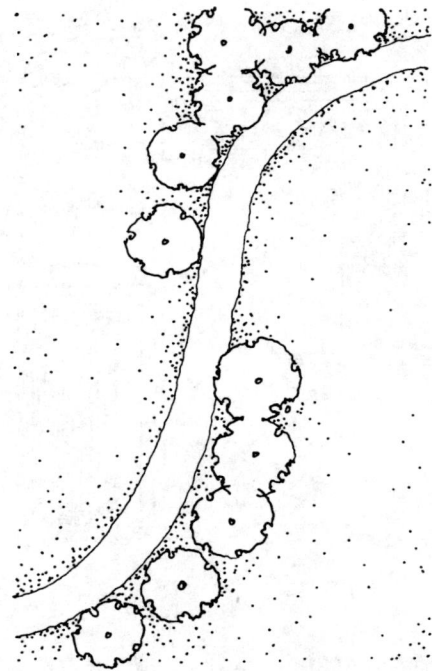

SHRUBS DEFINE A PATH

infinite variety of sizes, forms, textures, and colors of plants to choose from. In general, most of the plants in a site design should be more or less conventional and appropriate to their surroundings. However, plants of varied shapes, colors, and textures may be added to provide more interest. Too little variation is dull and monotonous, while too much is busy and even chaotic.

OTHER FACTORS

Pollution

All living things require sun, air, and water for life. Therefore, the quality of the air we breathe and the water we drink is of utmost importance. Neither air nor water is absolutely pure in nature; air contains dust, germs, and pollen, and water contains dissolved minerals and other impurities. But generally, these natural pollutants are controlled by natural processes; for example, air is cleansed and filtered by rain and trees.

PLANTS FRAME A VIEW

Man, however, has added a tremendous amount of pollutants to the air and water. Millions of tons of gaseous and solid matter, the combustion products of automobiles and industry, have been discharged into the atmosphere. And huge quantities of refuse of all kinds have been dumped into our lakes and rivers. As a result, air and water pollution have become critical problems.

Natural air movement can often mitigate air pollution. But sometimes, temperature inversion occurs, trapping air close to the earth's surface, so that pollutants in the air cannot escape until unstable air arrives. The result is the familiar pall of smoke, haze, and smog that hangs over our cities and industrial areas and even the countryside.

The site planner must be aware of possible pollutants on or near a site: Is the site downwind of an industrial plant or other source of air pollution? Is there a contaminated stream on the site? Or polluted ground water? Pollution, unfortunately, is a fact of today's life.

Noise

Every site has not only its own unique topography, soil conditions, and microclimate, but also its distinctive pattern of sounds. A place that makes us feel peaceful and serene is frequently one whose sounds evoke positive feelings in us, and these are most often the sounds of nature: running water, chirping birds, rustling leaves. In fact, so closely are such sounds associated with tranquility that they have been recorded, presumably permitting one to experience instant serenity, anywhere and anytime.

Man-made sounds, on the other hand, are often harsh and irritating, and are usually referred to as noise. Urban sites particularly abound with noise: your neighbor's stereo (never yours, of course), street traffic, garbage trucks, airplanes, etc. We attempt to control noise in much the same way that we block other undesirable elements, by utilizing barriers of one kind or another.

The loudness of a sound decreases as the square of the distance from a point source; in other words, doubling the distance from the source drops the sound level to one-fourth. Thus, distance is the simplest noise barrier.

Trees are somewhat effective against high frequency noise, but the most effective barriers are solid, such as structures, walls, or earth. The efficiency of a solid barrier varies with its height, the frequency of the sound, and the closeness of the barrier to either the source or the listener. Thus, a high wall close to the source is a good barrier against high frequency noise.

Therefore, there are several defenses against noise: first, we attempt to control it at its source. This is often difficult if not impossible, and so our next strategy is to place indoor or outdoor activities as far as possible from noise sources. Next, we shield activities from noise by using solid barriers. It is also possible to mask noise by adding desirable sound, such as water or music. The final protection against noise is the use of sound insulation in buildings, which is effective for indoor activities, but of course leaves the outdoors vulnerable.

Ecology

Ecology is defined as the study of the relations between organisms and their environment. While this may be an adequate definition for the biologist, it is too narrow for the site designer. In site design, ecology refers to an understanding of the interdependence between man and the physical environment, including plants and animals. In an urban environment, ecology takes on still a different meaning: the interdependence between man and the entire urban pattern, including circulation systems and cultural and social institutions.

The site planner must consider how the environment will be affected by his or her manipulation of site elements. Will the planned site development produce noise, or pollute air or water, or destroy valuable soil? Will natural patterns of drainage be upset, and if so, will the risk of flooding be increased? Will grading undermine adjacent streets or structures or be unsightly or increase the possibility of sliding? Will the planned site design modify existing traffic patterns, and will existing circulation systems be adequate? These and myriad other questions must be addressed and satisfactorily answered by the site planner, not merely out of a sense of responsibility and conscience, but also because frequently, an environmental impact statement must be filed and approved before any development is begun.

Thus, to some extent, the site planner has become the conscience of the community. He or she tries to leave existing points of interest, such as streams, rock outcroppings, and trees, undisturbed. The physical environment, including topography, is altered as little as possible. The development of marshlands and flood plains is limited to recreation and similar uses. Development in hazardous areas, such as those subject to sliding, subsidence, floods, or fires, is restricted.

The days of unlimited and uncontrolled development are over. Architects now recognize their responsibility not only to their clients, but also to the public and to the land itself.

LESSON 2 QUIZ

1. Inorganic silt which is designated MH in the Unified Soil Classification System has which of the following properties?

   A. High bearing capacity

   B. High compressibility

   C. Good drainage

   D. All of the above

2. Changing the composition of rock by chemical processes is known as

   A. decomposition.

   B. disintegration.

   C. weathering.

   D. erosion.

3. Select the correct statement.

   A. The upper several inches of soil, called the C horizon, is largely responsible for a soil's fertility.

   B. Erosion of the earth's surface has gone on at a constant rate for millions of years.

   C. Topsoil should be carefully conserved, since it has taken many centuries to produce.

   D. Humid climates tend to have gentle topography and shallow soil.

4. The system by which water on, under, and above the earth recirculates is called the water cycle. Select the correct statement about the water cycle.

   A. Most of the water that falls on the land as precipitation soaks into the ground, except in urban areas.

   B. Runoff consists of total precipitation less the water which infiltrates into the soil.

   C. Removal of vegetation increases the amount of runoff.

   D. Transpiration refers to the process by which water is evaporated from the oceans.

5. Which of the following statements about ground water are correct?

   I. Building foundations may not extend into the zone of saturation.

   II. A high ground water table requires that construction excavation be kept dry by pumping.

   III. The surface of underground water fluctuates seasonally, and the water flows very slowly.

   IV. Where the ground water table is high, providing waterproofing and gravel backfill relieves the hydrostatic pressure against basement walls.

   V. Since water seeks its own level, the ground water table is a horizontal surface.

   A. I, II, and IV   C. I, III, and IV

   B. II, III, and V   D. II and III

6. An underground permeable material through which water flows is called

    A. a ground water table.

    B. a soil horizon.

    C. a zone of saturation.

    D. an aquifer.

7. Select the INCORRECT statement about trees.

    A. They provide the most effective barrier against unwanted sound.

    B. They help to control erosion and flooding.

    C. They cool, humidify, and filter the air.

    D. They can provide both a vertical and a horizontal enclosure.

8. Select the most correct statement.

    A. Indigenous vegetation is not necessarily well suited to a site.

    B. The use of many exotic plant varieties adds to the naturalness and attractiveness of a site.

    C. In general, most of the plants in a site design should be more or less conventional and appropriate to their surroundings.

    D. Using plant materials to frame a view is artificial and should therefore be avoided.

9. Select the best definition of a flood plain.

    A. The area from which a river and its tributaries receive their water.

    B. The area bordering a stream over which water spreads when a flood occurs.

    C. The area bordering a stream which has been enlarged by levees to contain annual flooding.

    D. Any area formed as a result of recurrent floods.

10. Complete the following statement. In expansive soils,

    A. footings should be designed using a low bearing value.

    B. footings should not extend more than 18 inches below the surface.

    C. spaced foundation piers should be used, rather than footings.

    D. footings should extend below the depth of seasonal moisture change.

TOPOGRAPHIC ELEMENTS

INTRODUCTION

Topography

Topography is the graphic representation of an area's surface features. It is synonymous with landform or the shape of the ground. It may encompass mountains, rolling hills, prairies, and plains, while at a different scale, it may include mounds, ramps, berms, and even ripples in a sand dune. Topography has great environmental significance, since it affects the aesthetic character of an area, its microclimate, drainage, views, and the setting for structures.

Since the ground surface of every area has a unique configuration, the topographic map of a particular site will be unique as well. A topographic map allows one to understand the pattern of the land, since it indicates slopes, ridges, valleys, summits, stream beds, and drainage patterns. Topographic maps, therefore, are an invaluable aid for environmental assessment; in fact, topography is often the major determinant in site development.

Topographic information always influences and frequently determines site use, site circulation, distribution of utilities, placement of buildings, and the disposition of open spaces. A school playground, for example, requires a large, relatively flat area; a dead level area, on the other hand, may have drainage problems. Without a

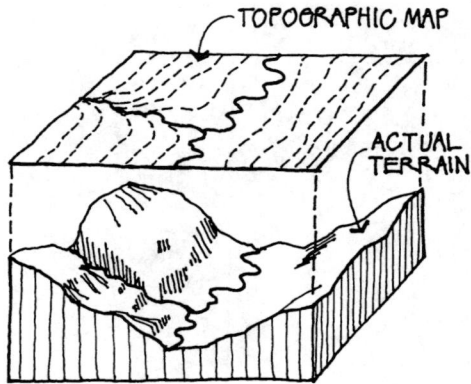

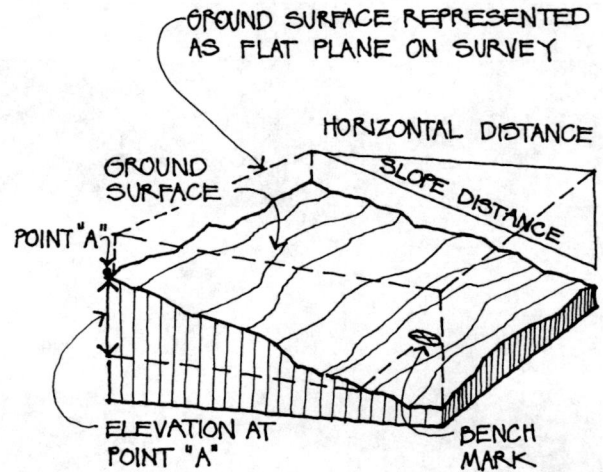

topographic map for reference, slope analysis for such problems would be impossible. Site design, therefore, is largely dependent on landform information provided by topographic maps.

Topographic Maps

Topographic maps are developed either from aerial photographs or surveys, where smaller parcels are involved. Aerial photos have the inherent advantage of depicting the land with a high degree of detail. Individual photos can be interpreted separately; however, they are most useful when combined in pairs and viewed stereoscopically. With the use of special equipment, aerial photos may be scanned to determine lines of equal ground elevation, that is, contours.

Topographic surveying consists of obtaining field data, from which a map is plotted showing the configuration of the ground surface. Regardless of the method used, the purpose of all topographic surveys is to produce a map on which variations in ground elevation can be readily observed and analyzed. Topographic maps also generally show property lines, roads, structures, trees, etc., in addition to ground surface elevations.

Surveys that extend over a relatively small area ignore the earth's curvature and assume the

ground surface to be a flat plane. Horizontal distances, therefore, are considered to be straight and are measured along a flat plane. Slope distances are never used for measurement. Vertical distances, or elevations, are designated as the distance above sea level or above any other bench mark, that is, any permanent point of known elevation. Both horizontal distances and elevations are measured and recorded in feet and hundredths of a foot, never in inches.

CONTOURS

Contour Lines

The shape of the ground surface is most often represented on drawings by contour lines. Contours are imaginary lines that connect all points of equal elevation. Each contour line may be thought of as the intersection of a level plane with the ground surface, such as a horizontal slice through a mountain or the shore line of a lake. Contours were developed in Holland in the mid-18th century, but it was not until the late 19th century that they became the common method for depicting terrain on survey maps.

The contour interval is the uniform difference in elevation between two adjacent contours.

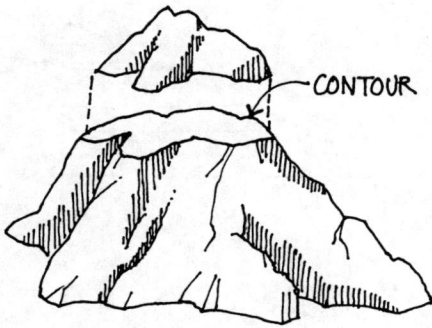

HORIZONTAL SLICE THROUGH A MOUNTAIN

SHORELINE IS ALSO A CONTOUR LINE

This interval is typically 1, 2, 5, or 10 feet, depending on the purpose and scale of the map and the character of the terrain represented. The contour interval is generally constant for the entire drawing and should always be indicated somewhere on the plan.

Contour Standards

Following are some standards that apply to the use of contours:

1. Existing contours are shown by a dashed line.

2. Every fifth contour is shown slightly darker for legibility.

3. Proposed (or modified) contours are shown by a solid line on the same drawing that shows existing contours.

4. Contour lines are labeled with the number within or on the high side of the line.

5. The contour interval is small for relatively flat areas, while for rough terrain, the contour interval is large.

6. The smaller the scale of the map, the larger the contour interval.

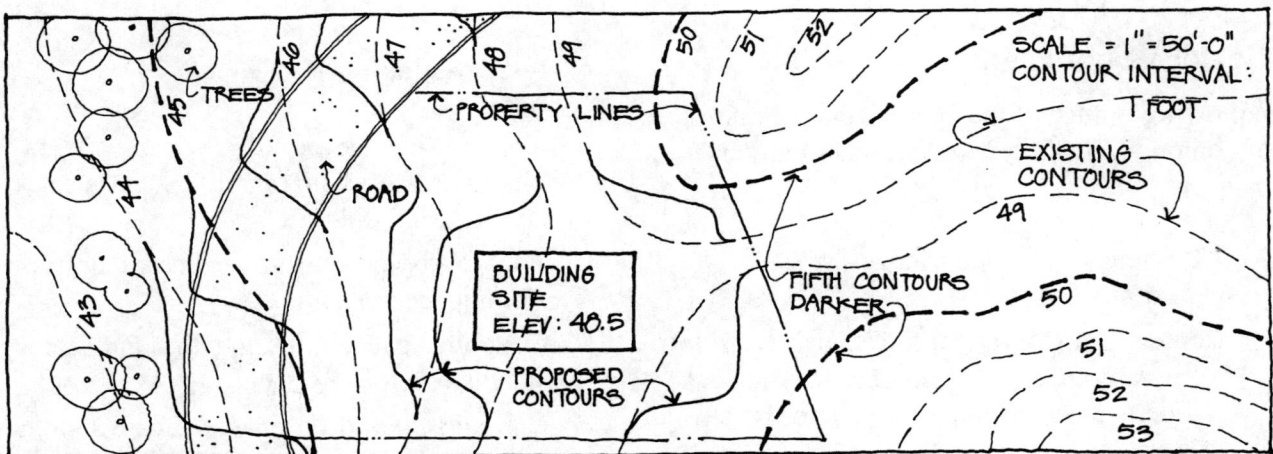

CONTOUR STANDARDS

Contour Patterns

For those who work with topographic maps, it is important to know the contour patterns of typical topographic features. Once these are mastered, it becomes a simple matter to understand the shape of a ground area and to readily modify landforms when required. Following are the most typical topographic forms:

1. *Uniform slopes* are indicated by parallel contours which are evenly spaced.

2. *Convex slopes* are shown by parallel contours spaced at increasing intervals going uphill. In other words, the closer contours are at the lower elevations.

3. *Concave slopes* are shown by parallel contours spaced at decreasing intervals going uphill. In this case, the closer contours are at the higher elevations.

4. *Valleys* are indicated by contours which point uphill.

5. *Ridges* are indicated by contours which point downhill.

6. *Summits* and *depressions* are represented by concentric closed contours. For both forms, spot elevations should be included at the highest or lowest point.

Contour Characteristics

Following is a list of the general characteristics of contours. The letters refer to the map on page 58.

A. All points on a contour have the same elevation.

B. Contour lines never split, although two identically numbered contours may appear side by side, at the top of a ridge or bottom of a valley.

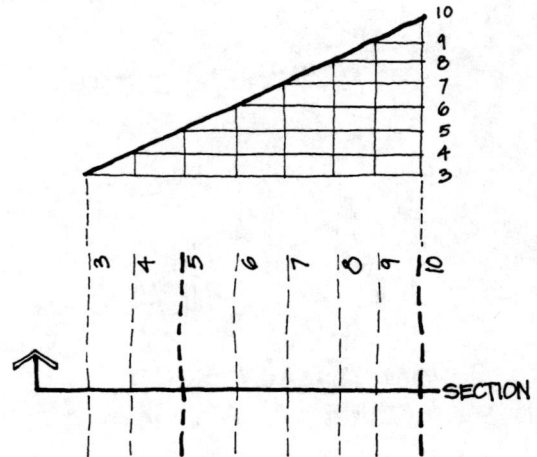

1 - UNIFORM SLOPE

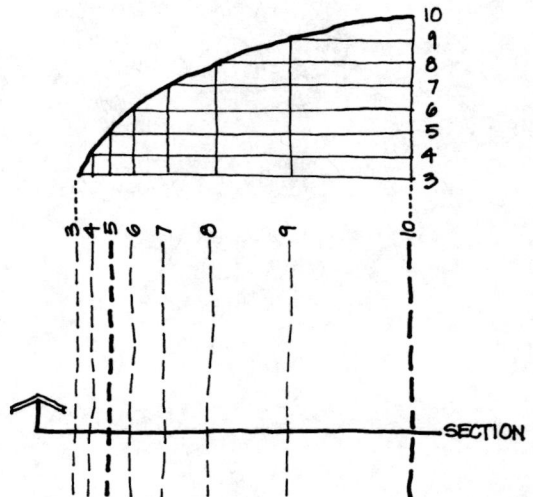

2 - CONVEX SLOPE

C. Contour lines never cross, except where there is an overhanging cliff, a cave, or similar configuration.

D. Equally spaced contours indicate a uniform sloping surface.

E. Contours spaced close together indicate a steep slope.

F. Contours spaced far apart indicate a slight grade.

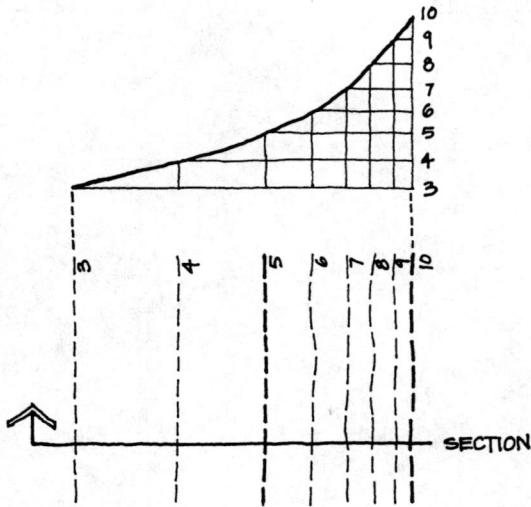

3 - CONCAVE SLOPE

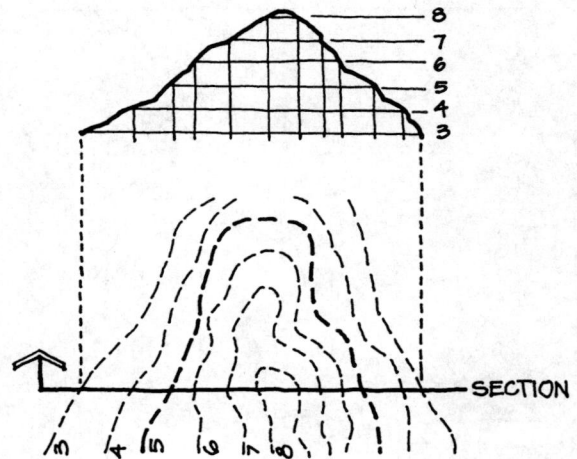

5 - RIDGE

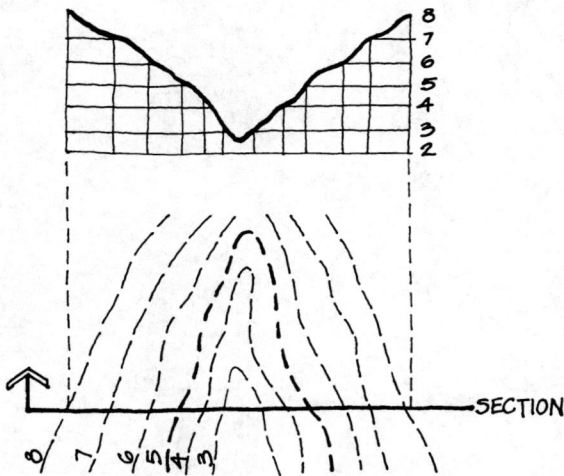

4 - VALLEY

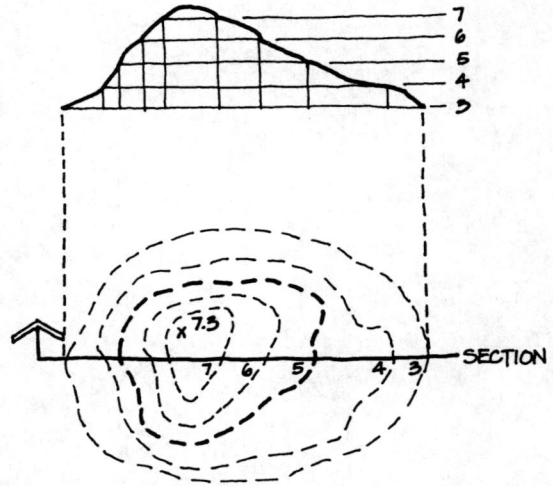

6 - SUMMIT

G. Contours spaced at increasing intervals (further apart) going uphill indicate a convex slope.

H. Contours spaced at decreasing intervals (closer together) going uphill indicate a concave slope.

I. Valleys are indicated by contours pointing uphill.

J. Ridges are indicated by contours pointing downhill.

K. A contour that closes on itself within the map area is either a summit or a depression.

L. Contours that run in straight parallel lines indicate a plane surface.

M. Drainage always occurs perpendicular to the contours, because this is the shortest distance and hence the steepest route of travel.

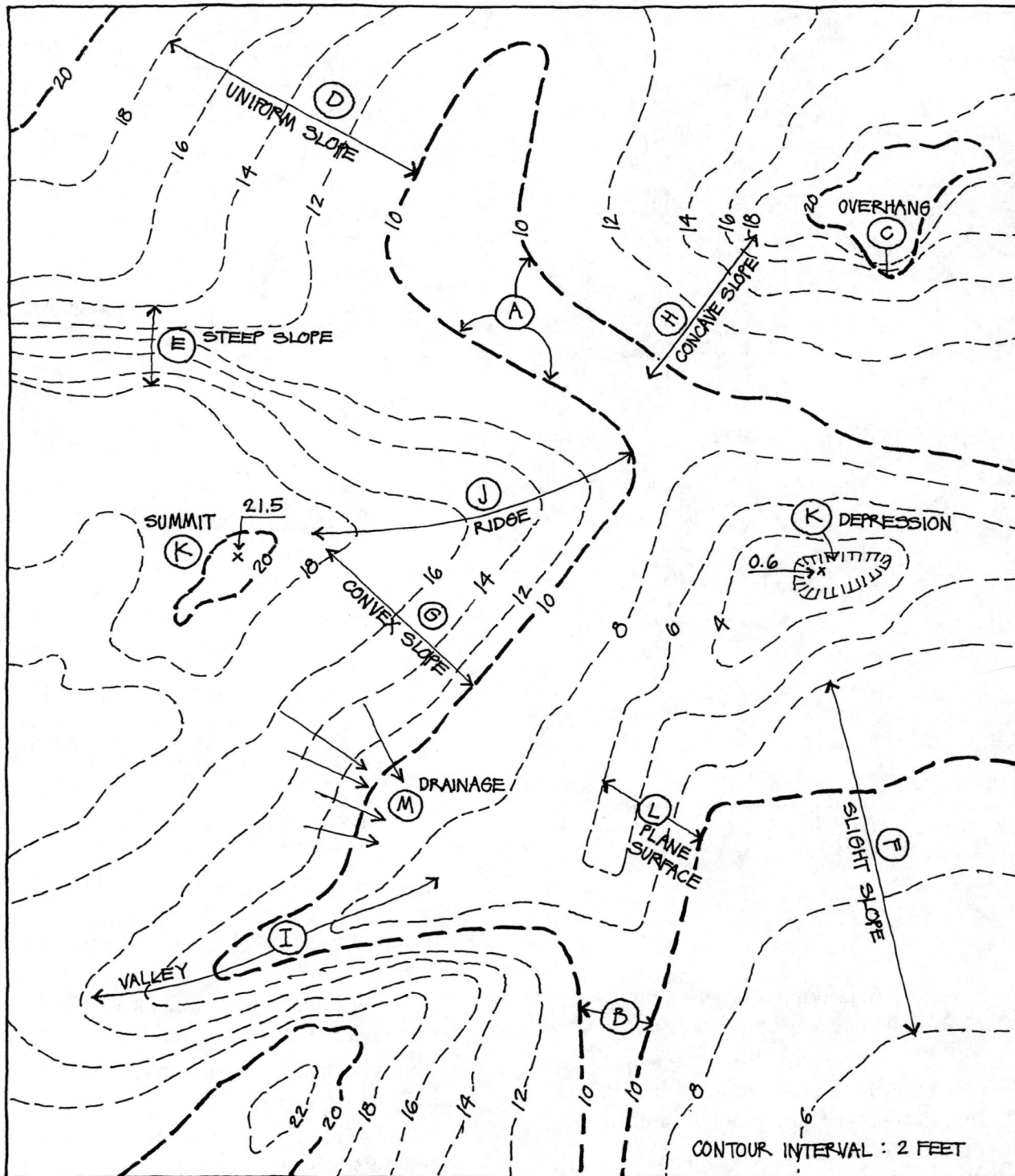

CONTOUR INTERVAL : 2 FEET

Representing Topography

Although contours are the most common way of representing topography, there are other methods used to depict topographic relief. Among these are the following:

Spot elevations. A spot elevation is a number corresponding to the exact elevation at a key point on the ground. Spot elevations are designated by an arrow pointing to the exact spot where the elevation is located, and they are used to indicate high or low points, tops of curbs, bottoms of walls, bases of trees, floor levels of structures, building corners, etc. Spot elevations may also be used on a uniform grid with precise elevations noted at every intersection point of the grid. Contours may be determined from spot elevation grids through interpolation.

Shading. There are several shading methods one can use to depict ground form, and all of them involve shading the slopes in proportion to their degree of steepness. Shallow slopes are lighter with the shading lines further apart, while steep slopes are made darker by placing the shading lines closer together. The most common

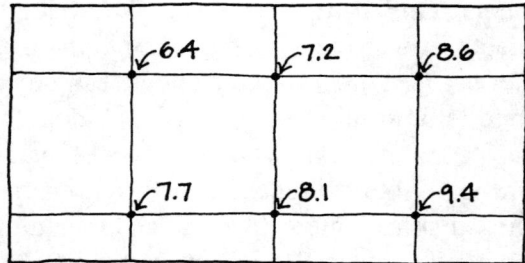

SPOT ELEVATIONS ON GRID

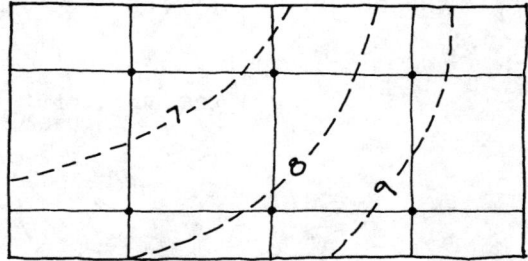

CONTOURS INTERPOLATED FROM SPOT ELEVATIONS

shading system employs *hachures*, which are short, disconnected lines drawn perpendicular to the slope, or in the direction of water flow. Hachuring requires knowledge of the area's contours, since the shading lines are drawn between two successive contour lines. Hachured maps give a vivid picture of the ground terrain, but it is difficult to determine exact elevations unless spot grades are indicated. For this reason, hachuring is rarely used today.

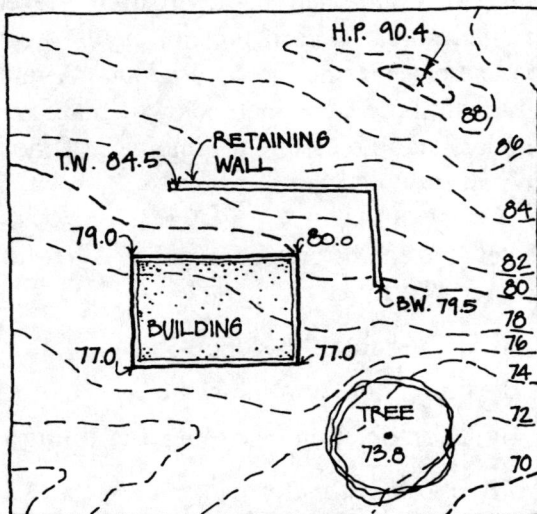

SPOT ELEVATIONS

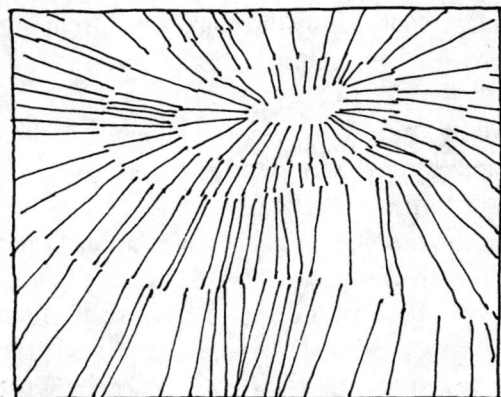

HACHURING

Models. One of the best ways of studying the general form of the terrain is through the use of relief scale models. Contour plans can be converted into useful three-dimensional models by transferring the contours onto a layer of material (cardboard, polystyrene, etc.) whose thickness represents the contour interval, and then gluing the layers together. Rough models are quickly and easily made, and they enable the viewer to comprehend the ground form of an entire site at a glance.

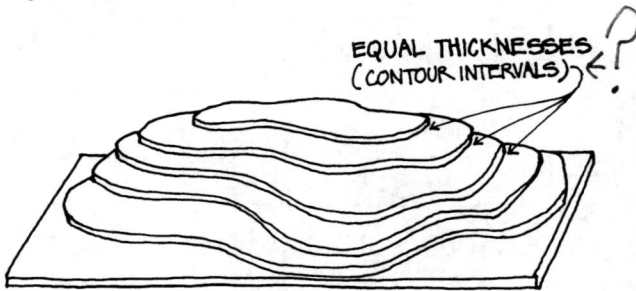

RELIEF SCALE MODEL

GRADING

Cut and Fill

The development of a site almost always requires a certain amount of topographic modification, that is, grading. In general, grading should be kept to a minimum, so as not to upset existing drainage patterns, disturb native landscaping, or destroy an area's natural aesthetic value.

There are only two ways to change the land surface: by removing earth (cutting) or by adding earth (filling). The greatest economy results when the amount of earth cut is approximately equal to the amount filled. Grading plans indicate cut or fill by means of new, solid contour lines, which represent proposed modifications to the existing contours, which are shown dashed. A proposed contour that moves in the direction of a lower contour line indicates fill,

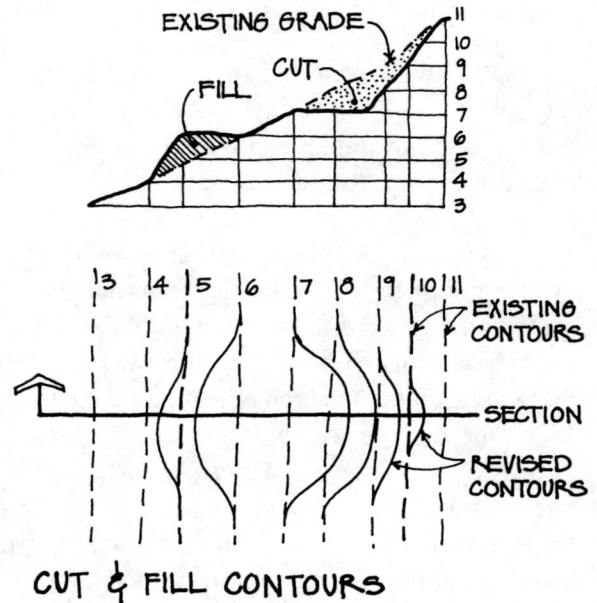

CUT & FILL CONTOURS

that is, the addition of earth. Conversely, a proposed contour that moves in the direction of a higher contour line indicates cut, or earth removed. All grading involves cutting, filling, or a combination of the two.

Level Areas

Almost all structures require a fairly level ground area on which to sit. Creating a level area, therefore, represents one of the most common exercises in modifying contours. A familiarity with these basic methods will enable one to understand the essential principles of modifying all ground form.

A level area can be created in one of the following three ways:

1. By cutting into the slope
2. By filling out from the slope
3. By a combination of cutting and filling

Cutting into the slope provides a level area by essentially scooping it out of the hillside. You begin by locating the area to be leveled on the topographic map. The elevation of the contour just downhill of the level area will be the finish grade for the entire level area. Next, you take the nearest higher contour and wrap it around the high side of the level area and connect it to the same existing contour. This first modified contour should be located a few feet away from the level area to allow for the ground slope. For example, if the contour interval is one foot and the finish grade of the slope is to be 2:1 (2 feet horizontal to 1 foot vertical), then the first modified contour should be drawn two feet away from the level area. Proceed now to the next higher contour and wrap it around the level area in the same way, keeping the same distance between contours, that is, two feet for a 2:1 slope. Continue up the hillside until there are no more overlapping contours, being certain that the two-foot minimum distance between contours is maintained.

It is important to select a gradient for the new sloped bank that is steeper than the existing grade, otherwise you will never be able to meet the existing grade, that is, the excavation of earth will never end. Normally, one attempts to minimize grading by meeting the existing grade as quickly as possible, provided that the new bank is a stable slope. With steeper slopes, vertical cuts and retaining walls may be necessary.

Filling out from the slope is substantially the reverse of the method just described. Instead of scooping out the hillside, you create a level shelf by adding earth. Again you begin by locating on the topo map the area to be leveled. The finish grade of this area is set at the same elevation as the contour immediately uphill of the level area. Next, you take the nearest lower contour and wrap it around the low side of the level area. As before, the contour is spaced far enough away from the level area to accommodate the desired slope between contour lines. You continue downhill, wrapping contours around the level area, until your new slope angle meets the existing grade.

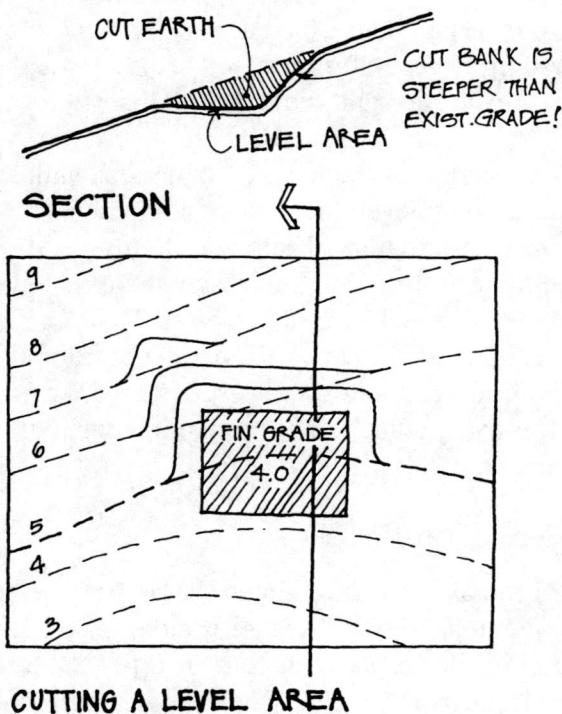

SECTION

CUTTING A LEVEL AREA

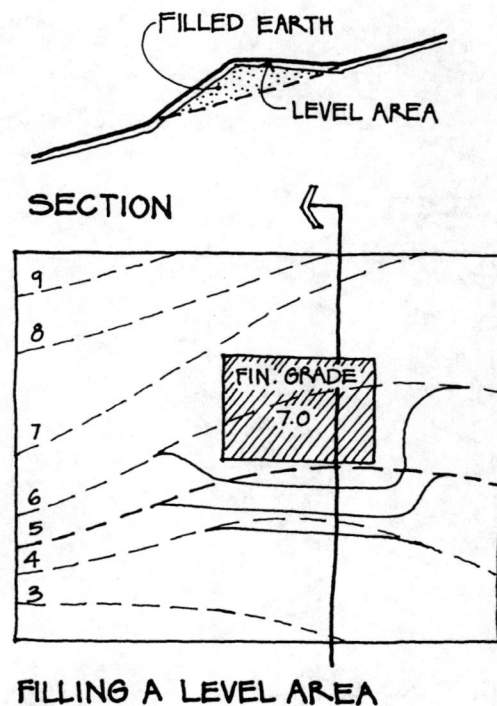

SECTION

FILLING A LEVEL AREA

A combination of cutting and filling is the most common grading method for providing a level area, since this method attempts to balance cut and fill. If the amount of earth cut is equal to that used for fill, the need to import or dispose of soil is eliminated and grading costs are reduced to a minimum. After locating the area to be leveled, you establish the area's finish grade by determining the existing grade at its midpoint, which is at the intersection of the level area's diagonals. Next, you wrap the adjacent contour above the level area around the high side, and the next lower contour around the low side. This same procedure is followed, as before, until all contour lines requiring modification have been adjusted.

Given a choice, which grading method should be used? Often, the choice is out of the designer's hands. For example, filling operations may not be feasible on steep sites, in which case cutting may be the only solution. But, cut earth must be hauled away, which is expensive. Cut earth is generally more stable than filled

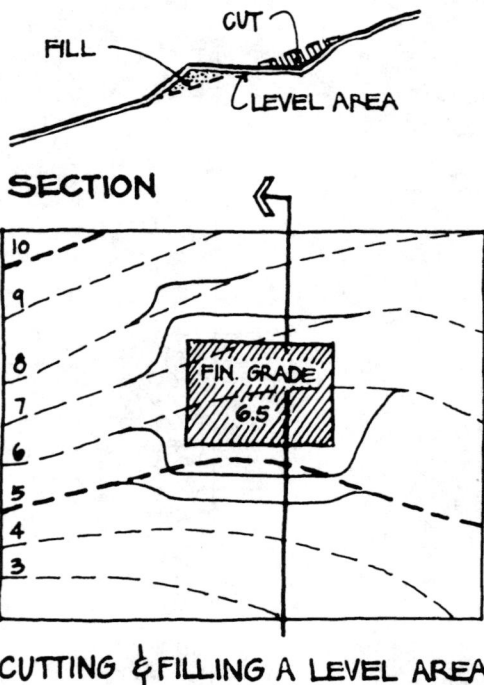

SECTION

CUTTING & FILLING A LEVEL AREA

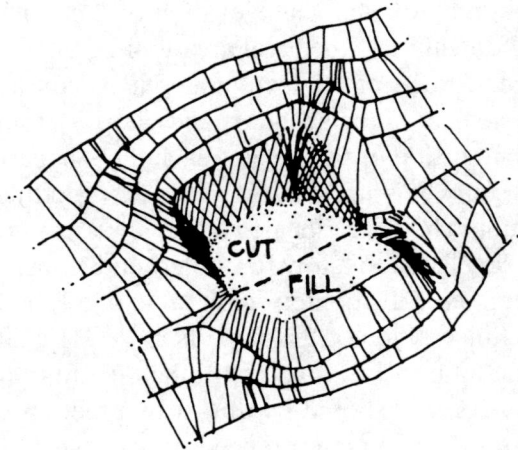

CUT & FILLED LEVEL AREA

earth, and therefore, cut slopes are generally permitted to be steeper than fill slopes.

Filled earth may be used where relatively flat, low areas must be raised to make them usable. However, filled earth has several drawbacks: it is not always suitable to support new construction, it often requires expensive compaction, and it is frequently subject to slippage and erosion. If one has a choice, the cut and fill method of grading is generally preferred because it is the least expensive and the most flexible.

The location of the level building area and, to some extent, even the area's finish grade is somewhat arbitrary. There are often several solutions to the design of a site and one may have to proceed by trial and error. Therefore, if grading does not result in a perfect solution the first time (precipitous slopes, destroyed trees, excessive grading, etc.), one should remain flexible and try again.

Circulation Paths

In general, circulation paths should be as level as possible, simply because it takes less effort to move horizontally than up a slope. Natural grade, however, is almost never level; therefore,

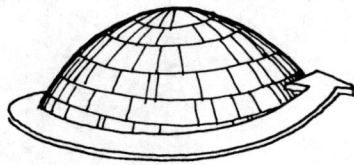

GRADUAL DUT
LONGEST ROUTE

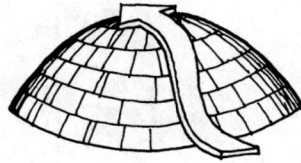

QUICKEST AND
SHORTEST ROUTE

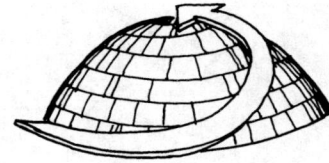

MOST NORMAL
ROUTE

CIRCULATION PATHS

circulation invariably requires sloped paths as well as level ones. In selecting a route for circulation, one can move parallel to the contours, which is level but requires extensive grading because of its length, or perpendicular to the contours, which results in the steepest path but requires less grading. Normally, circulation paths are graded somewhere between perpendicular and parallel to the contours.

Before laying out a path, one must determine its required width and the maximum allowable steepness. In addition, the slope of the path should be as uniform as possible. This means that contour lines along the path's route will be more or less evenly spaced. It is usually best to locate paths in valleys or along stream beds and avoid placing them on steep slopes. Highways commonly run long distances around a mountain, rather than going directly over or through it and creating extensive grading or tunneling problems.

Grading circulation paths is accomplished the same way as creating level areas, that is, one can cut, fill, or use a combination of cut and fill.

Cutting a path in the slope requires first locating the path to scale on the topographic map. Next, find the intersections of the contours with the low side of the path and extend each contour perpendicularly across the path. When you reach the other side, run next to the path until each

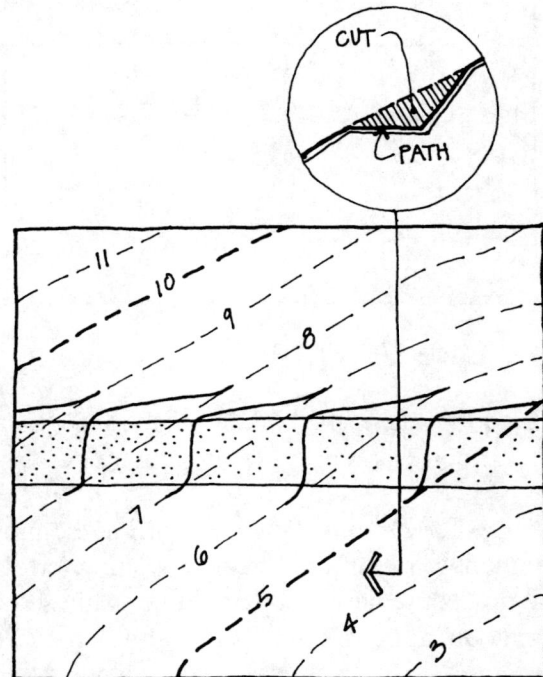

CUTTING A PATH

contour connects back to its corresponding existing contour line. This will create a graded bank on the uphill side of the path, the steepness of which can be modified through contour manipulation.

Filling to create a path works in the reverse way. After laying out the proposed path in plan, locate the intersections of the contours with the path's high side. As before, cross the path perpendicularly with each contour, run them next

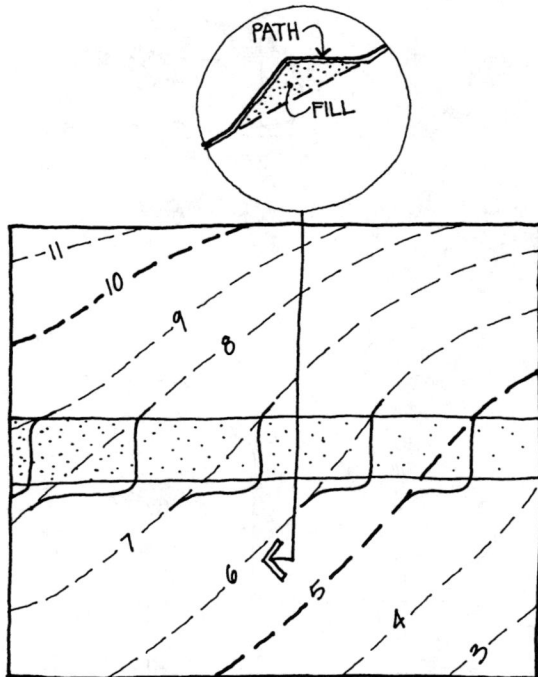

FILLING A PATH

to the low side of the path, and reconnect them to their corresponding existing contours.

Using a combination of cut and fill is the most common way to grade a path on sloping terrain. The procedure here is to lay out the path, select a contour in the middle, establish a perpendicular contour that crosses it at its midpoint, and then reconnect the contours on both sides. By so doing, you will create a cut bank above the path and a filled bank below it.

If the natural grade is not excessively steep, paths may be designed to run perpendicular to the contours, that is, straight up the slope. After the path is located on the topographic plan, the path's steepness must be established by evenly spacing contours perpendicularly along the path's route. It is best to begin in the middle of the path's length and allow that contour to run right through undisturbed. There may be times, however, when this is not possible, and you may have to begin

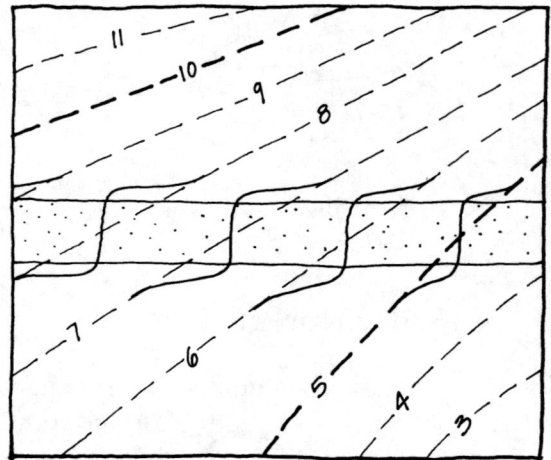

CUTTING & FILLING A PATH

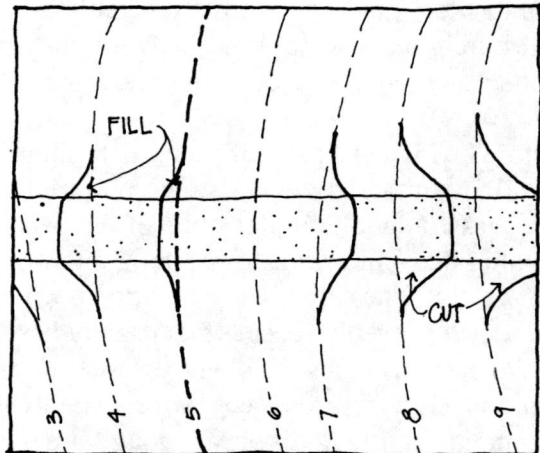

PERPENDICULAR PATH

with an established elevation at one or both ends of the path. The result of grading a path perpendicular to the contours may be either cut banks or filled banks on both sides of the path.

The Grading Process

The grading process begins with the development of a grading plan. This plan shows the site boundaries, existing topography and site features, and the proposed modifications, including new structures. Grading plans use contours, notes, and a variety of special symbols to describe what must be done to reshape the land into the desired forms.

The principal aim of all grading work is to make the land appropriate for its purpose and to preserve a stable system. The finished grade should have positive drainage, stable slopes, balanced cut and fill, and pleasant and harmonious visual forms.

Grading operations begin by removing the topsoil, which is stored on the site and later reused over the modified ground forms. Grade stakes are then placed at intervals in the subsoil to indicate the required new levels. Grade stakes are located at all critical points, such as peaks, valleys, roads, walls, and other points of grade change. Grading machines then cut or fill the earth to the staked levels and shape it into the desired configuration. Among the machines employed for this purpose are bulldozers, scrapers, graders, power shovels, rollers, and scarifiers. With the extensive range of machinery available today it is rarely necessary to do any expensive hand shoveling.

The following summarizes the most important general rules of grading:

1. Do not extend grading beyond the property lines.
2. Strip and save all topsoil prior to grading.
3. Avoid the destruction of valuable existing vegetation.
4. Attempt to balance all cut and fill.
5. Avoid flat grades that create drainage pockets.
6. Avoid erosion by grading slopes within their natural angle of repose.
7. Be certain finish grades enable water to flow away from all structures.
8. Avoid grading solutions that rely on expensive retaining walls, steps, or other construction.

There are several serious hazards associated with grading that may cause irreversible damage. Some of these are as follows:

1. *Loss of topsoil.* It takes hundreds of years of natural decomposition of organic material to produce a thin layer of topsoil. Since it is absolutely essential to plant life, topsoil must be retained.
2. *Loss of vegetation.* Vegetation replenishes the oxygen, moderates the climate, and helps control erosion. All mature plants, therefore, must be preserved.
3. *Altered drainage patterns.* Modified runoff patterns can cause erosion and contamination of downstream waterways. New drainage patterns must be carefully planned.
4. *Unstable earth.* Grading earth that is unstable can produce slides, slippage, and cave-ins. Work done at one location can affect other sites, even if relatively distant.
5. *Aesthetic damage.* Grading that alters existing site qualities may destroy the uniqueness of an entire area. Designers should remain sensitive to existing conditions.
6. *Unique conditions.* Some areas are excessively steep or contain extensive rock outcroppings; others provide a natural habitat for wildlife. Such areas, if possible, should be left in their natural state and not graded or developed in any way.

Although grading is sometimes considered a mechanical process, the site designer who creates a grading plan is actually modeling earth forms in much the same way as an artist sculptures clay. Sensitive grading represents the best transition between the existing land and the proposed requirements of the project. It should be functional, economical, and attractive, and do a minimum amount of ecological damage.

DRAINAGE

Storm Drainage Systems

Drainage of land refers to the methods used to collect, conduct, and dispose of unwanted rain water. During a rainstorm, water is absorbed by the soil in varying amounts and at varying speeds of percolation, depending on the soil's porosity and the volume of soil above the ground water table (see page 45). If rain falls faster than the soil is able to absorb it, puddles form and surface water begins to flow downhill. The uncontrolled flow of this surface runoff may result in erosion, flooding, or damage to landscape and structures.

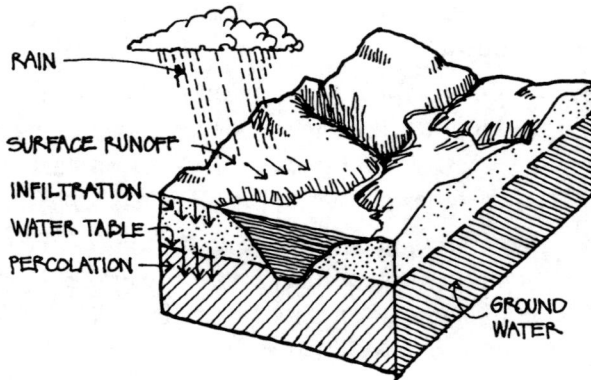

NATURAL DRAINAGE

Storm drainage systems have as their objective the removal of excess rain water. They are a substitute for natural surface drainage and are generally necessary only where development is dense enough to cause excess runoff following a rain. Low-density developments often rely on natural open land, but higher densities imply more paved areas, which cause greater runoff.

A typical drainage system begins with the roof water from an individual building. This water flows to roof drains and downspouts that eventually conduct it to the street, either directly or over paved areas or lawns. Once in the street,

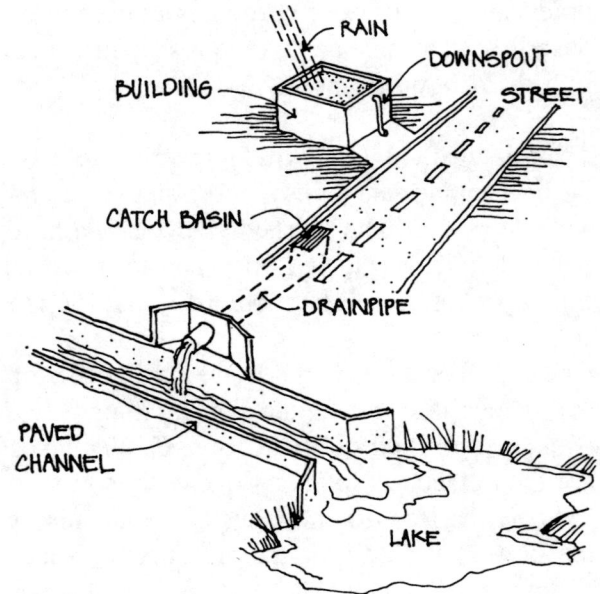

STORM DRAINAGE SYSTEM

the water flows downhill until it reaches a catch basin, where it continues in an underground drain line. The drain line may lead to a concrete channel that ultimately discharges the original roof water, plus all other runoff, into a lake or other body of water.

Storm drainage systems are designed to achieve the following:

1. Reduce flooding and damage by eliminating excess rain water.

2. Reduce erosion by controlling the rate and volume of water flow.

3. Eliminate standing water which may lead to pollution and insects.

4. Enhance plant growth by reducing soil saturation.

5. Improve the load-bearing capacity of soils.

Underground storm drainage systems are expensive, therefore, every effort should be made to employ surface systems, such as gutters, ditches, channels, short culverts, and especially, spacious planting areas.

Drainage Needs

The demand for drainage is determined by the following factors:

1. *Topography.* Steep areas drain quickly, often too fast to allow for percolation. Therefore, drainage benches or channels should be provided above and below steep banks to collect runoff water.

2. *Type of soil.* Soil type determines the amount and speed of water absorption, which in turn affects the runoff quantity.

3. *Vegetation.* Thick ground cover slows down the rate of runoff, reduces erosion, and reduces the need for elaborate drainage systems.

4. *Rainfall data.* Local rainfall data is necessary to calculate the frequency and intensity of rain water to be drained.

5. *Land use.* In rural areas, one can generally allow water to disperse over the landscape. In urban areas, on the other hand, surface runoff occurs for short distances only and then must be directed to sub-surface drainage devices.

6. *Size of area.* The area to be drained generally refers to those areas with limited percolation, such as roofs, roads, driveways, etc. Fields and open spaces can usually remain wetter than parking areas and paved walks.

Accurate runoff flows can be computed by use of the Rational Formula, which is $Q = CIA$, where Q is the quantity of runoff in cubic feet per second, C is the coefficient of runoff, I is the rainfall intensity in inches per hour, and A is the drainage area in acres. The coefficient C is the fraction of total rainfall that runs off a surface. Roofs and pavements have a value of

about 0.9, while lawns and other planted areas have a value around 0.2.

Surface Drainage

Surface drainage involves the removal of runoff water by means of surface devices only. It is generally preferred over subsurface systems, because it is less expensive, it allows some water to percolate into the ground, and the danger of clogged pipes is eliminated. The following are some general rules of surface drainage:

1. Water flows as a result of gravity, therefore, all surfaces must be sloped for drainage.

2. Water always flows perpendicular to the contours.

3. Good drainage requires a continuous flow. Slow-moving water may create bogs, while water moving too fast will cause erosion.

4. Water should always be drained away from structures.

5. Large amounts of water should never be drained across circulation paths.

6. Desirable slopes for surface drainage are as follows:

    Open land - 1/2 percent min.

    Streets - 1/2 percent min.

    Planted areas - 1 percent min. to 25 percent max.

    Large paved areas - 1 percent min.

    Land adjacent to buildings - 2 percent min.

    Drainage swales - 2 percent min. to 10 percent max.

    Planted banks - up to 50 percent max.

Following are the most common methods of surface drainage:

1. *Swale.* Sloping areas can be drained by creating swales, which are graded flow paths similar to valleys. Swales are graded around structures with finish contours always

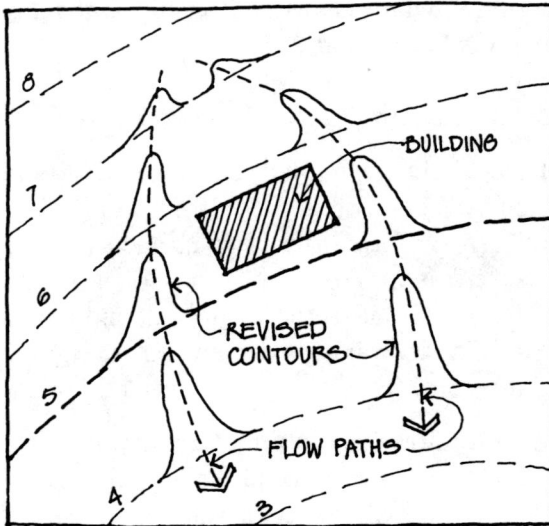

CREATING SWALES

pointing uphill and flow paths shown perpendicular to the revised contours.

2. *Sloping plane.* This is the simplest, cheapest, and consequently, the most common way to drain a relatively level area. The area tilts in one direction, so that the water drains to the low side. Adjacent structures are always located at the high side.

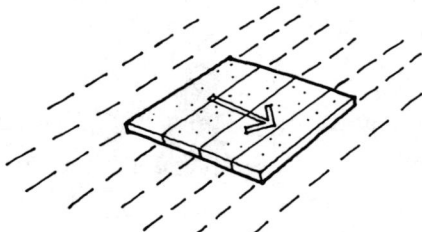

SLOPING PLANE

3. *Warped plane.* The high side is level, similar to the sloping plane. The contours, however, are fan-shaped, so that the entire area drains to one low corner.

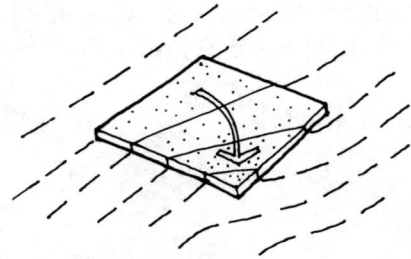

WARPED PLANE

4. *Gutter.* Gutters are formed by two sloping planes that create a valley. The planes are slightly warped so that water can run down the valley to a collection point. When adjacent to a structure, the top edge of one sloping plane will be level.

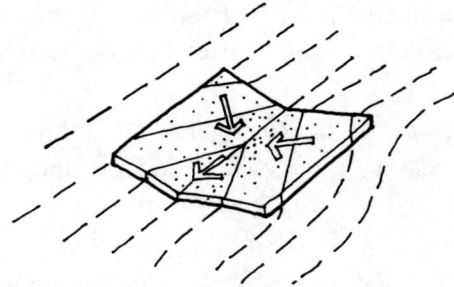

GUTTER

5. *Central inlet.* Large flat areas, especially where enclosed (courtyards, patios, etc.) employ a central drain toward which all surfaces slope. The disadvantage of this arrangement is that it requires a catch basin and sub-surface piping to dispose of the water.

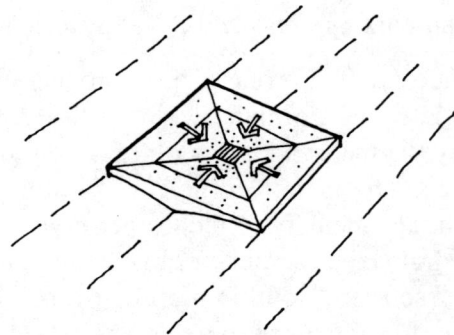

CENTRAL INLET

Sub-Surface Drainage

Sub-surface drainage refers to the collection, conduction, and disposal of water below ground level. Runoff water flows not only on the surface, but also below the ground surface. Collection of sub-surface water utilizes gravel-filled ditches and perforated drain pipe, or drain pipe laid with open joints. Runoff water that seeps into the earth flows vertically through the gravel until it reaches the openings of the drain pipe. This pipe collects the free-flowing water and carries it away in the direction of the sloping pipe.

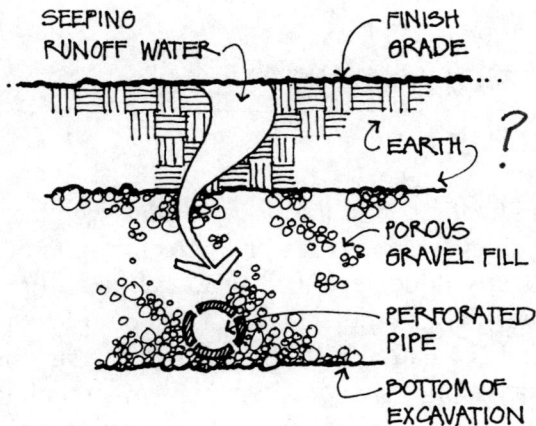

SUB·SURFACE DRAINAGE

Closed sub-surface systems consist of various fabricated collectors together with sections of closed pipe, which are used to carry water below grade from collection points to disposal areas. Such systems are useful in level areas, since drain lines can be buried progressively deeper, assuring an adequate slope for drainage.

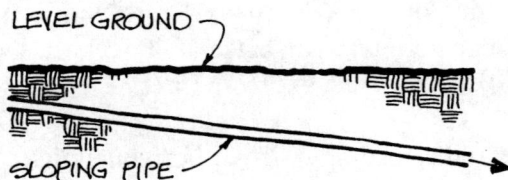

DRAINAGE INDEPENDENT OF GROUND SLOPE

Among the common fabricated collectors are the following:

Area drain. This device collects water from the low point of a limited area and conducts it directly to underground pipes. It has a metal grate to prevent debris from entering and clogging pipes.

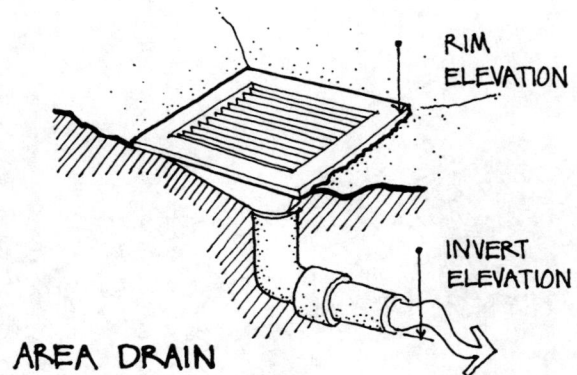

AREA DRAIN

Catch basin. This is similar to an area drain, except deeper and generally larger in order to catch and retain sediment which may clog the system.

Trench drain. This device is used to collect water along a wide strip before conducting it to underground pipes. It is suitable at the entrance to an underground garage, for example, where it collects the runoff water flowing down a sloping driveway.

Drainage pipes are manufactured from clay, concrete, plastic, or composition materials. They are rarely less than four inches in diameter and require a minimum slope of one percent to assure proper flow. On drawings, the depth of a drain pipe is indicated by the invert elevation, which is the elevation of the bottom of the pipe leading out of the drainage collecting device. The top of the device is indicated by the rim elevation.

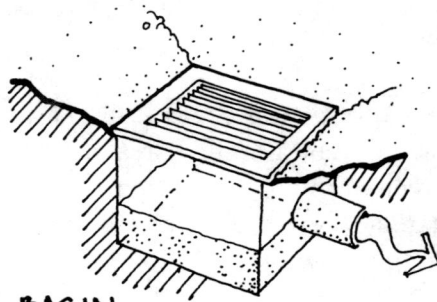
CATCH BASIN

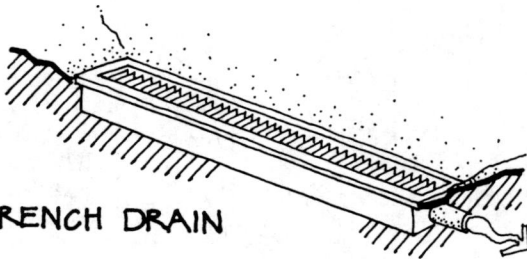
TRENCH DRAIN

Pipes which run underground beneath roads, driveways, or paths are referred to as *culverts*. They vary in size from six inches to several feet. Culverts should be straight, cross the road at right angles, and be sufficiently strong to resist moving traffic loads.

CULVERT

Following are some sub-surface drainage general rules:

1. Always determine a site's disposal points before designing the drainage system.

2. Sub-surface drain lines should travel in straight lines; changes of direction occur only at catch basins.

3. Avoid running drain lines beneath or through building foundation walls, retaining walls, or other construction.

4. Drain lines should follow the natural site slope as much as possible to minimize the depth of trenches.

5. Drainage systems often require several lines, and a branching system is frequently the most efficient solution.

DESIGN METHODS

Gradients

Gradient refers to the slope of land, which may be expressed as a ratio or a percentage, or occasionally in degrees. It is useful to know the gradient of a landform for a number of reasons: this information is used in slope analysis, maximum slopes are specified this way in building codes, and different soils have varying angles of repose, that is, the steepness beyond which soil will slide.

Gradient ratios express slope as the ratio between a horizontal distance and the vertical elevation change within that distance. By convention, the first figure of the ratio represents the horizontal distance, while the second figure (reduced to a factor of 1) represents the vertical elevation distance. For example, if land rises 20 feet in a horizontal distance of 40 feet, the slope is 40:20, which is expressed as 2:1.

As another example, if the horizontal distance between the 6 foot and 10 foot contours on a topographic map scales 12 feet, the slope is 12:4 (10 minus 6), which is expressed as 3:1.

2:1 SLOPE

3:1 SLOPE

33% SLOPE

Likewise, when scaled from a topographic map, if the horizontal distance between the 24 and 28 foot contours is 10 feet, the slope is 4/10 = .40, which is expressed as 40 percent.

40% SLOPE

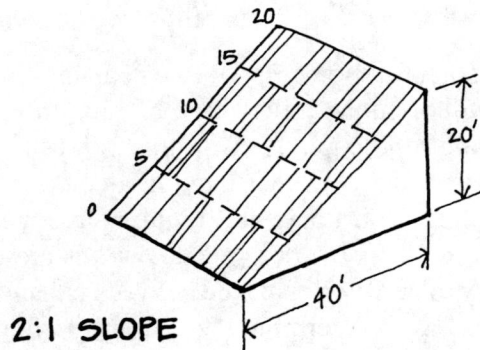

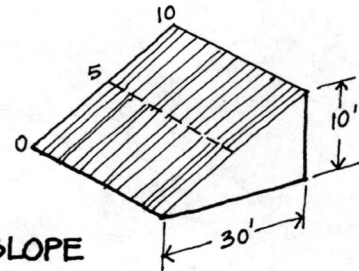

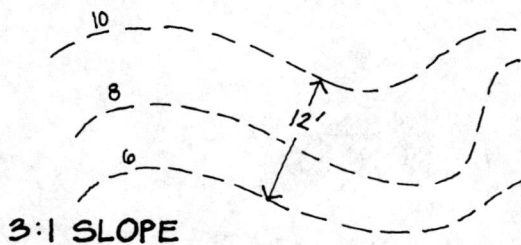

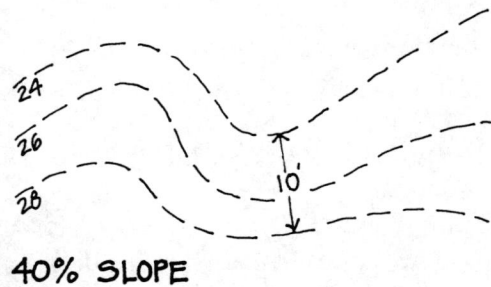

Gradient ratios are generally used for slopes on small-scale projects, and also to express design standards, such as:

2:1 - maximum allowable ground slope with little danger of erosion

3:1 - maximum slope for most planted areas

4:1 - maximum slope maintainable with a lawn mower

Gradient percentages express slope as the gradient (G) obtained by dividing the vertical elevation change (V) by the horizontal distance (H) in which the change takes place, expressed as a percentage. Therefore, G = V/H, where

G is the gradient in percentage

V is the vertical rise

H is the horizontal run

For example, if land rises 10 feet in a distance of 30 feet, the slope is 10/30 = .33, which is expressed as 33 percent.

The percentage method is more widely used than the ratio method, and it is also used to express design standards, such as:

0–1% - Flat slope, poorly drained, generally undesirable for development.

1–5% - Considered ideal for most development.

5–10% - Suitable for most development, maximum for walks.

10–15% - Considered too steep for many land uses without grading.

15%+ - Too steep without substantial development efforts and costs.

The percentage method may be used to calculate the vertical rise or horizontal run when one or the other of these values and the gradient are known. For example, if the gradient of a road is 6 percent and the elevation at point A is 21.7,

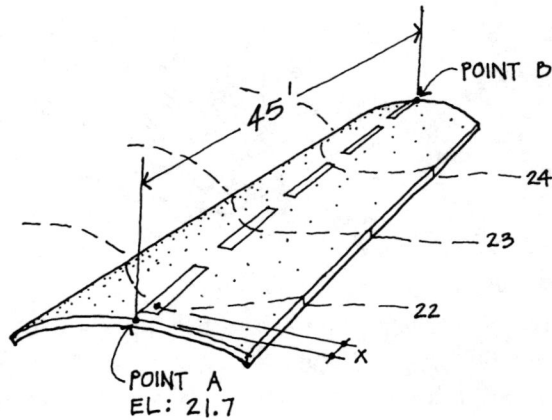

ROAD WITH 6% GRADIENT

then the horizontal distance X from point A to contour 22 is calculated as follows:

$X = H = V/G$ or $H = (22–21.7)/0.06 = 0.3/0.06 = 5$ feet

Similarly we can calculate the elevation at point B if the distance between points A and B is 45 feet, as follows:

$V = GH$ or $V = 0.06(45) + 21.7 = 2.7 + 21.7 = 24.4$

Percent of slope must never be confused with angle of slope. For example, a 45-degree slope, which is very steep, has a slope ratio of 1:1 and a slope percentage of 100 percent.

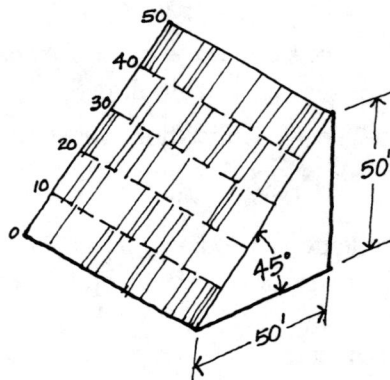

45° - 100% - 1:1 SLOPE

Calculating Cut and Fill

It is sometimes necessary to calculate the amount of cut and fill caused by modifying the existing topography. It is at this point that designers discover that moving a contour line half an inch on a plan may require moving a ton of earth 50 feet. Grading calculations are necessary to verify the balance between cut and fill, as well as to determine the cost of grading.

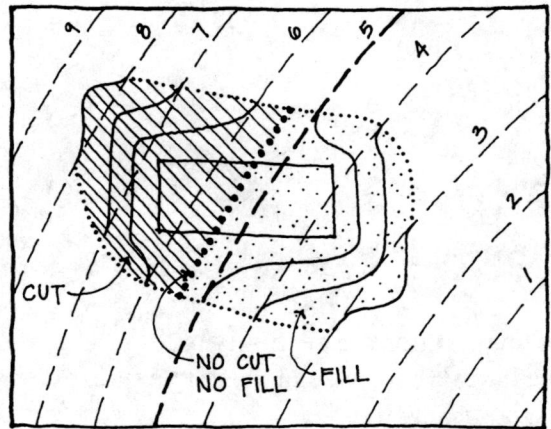

CUT AND FILL

Cut and fill quantities are expressed in volume, generally cubic yards of earth, and they are calculated as width × length × depth. The problem with calculating volumes of earth, however, is that contour shapes are almost always irregular. Therefore, quick calculations will produce only approximate results.

To determine the volume of cut or fill, it is necessary to calculate the area of a particular contour's horizontal surface (width × length) and then multiply it by the contour interval (depth). This can be done in two different ways. First, the irregular area can be divided by equally spaced parallel lines drawn a convenient distance (d) apart. Starting at one end, you measure the various lengths of these lines (L_1, L_2, etc.) and multiply the average length of two adjacent lines by the width d. The sum of all these areas

$$\left(\frac{L_1 + L_2}{2}\right) \times d + \left(\frac{L_2 + L_3}{2}\right) \times d +$$

etc. will be the total area of the irregular form. This area is then multiplied by the contour interval to obtain the volume.

The second way to compute the area is to overlay the irregular form with transparent graph paper, preferably with the grid paper at the same scale as the drawing. In that case, since each square represents one square foot, you simply count the number of squares within the irregular form to obtain the area. The area is then multiplied, as before, by the contour interval to obtain the volume.

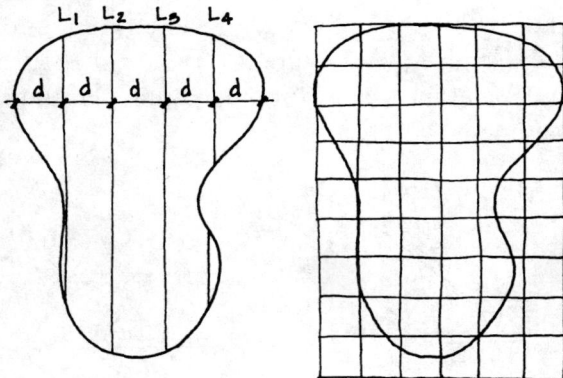

STRIP METHOD   GRAPH METHOD

AREA OF IRREGULAR SHAPES

After each contour volume has been calculated, you add up the cut and fill volumes to derive the total cubic footage of earth for each. These totals are then converted to cubic yards by dividing by 27, which is the number of cubic feet in a cubic yard. The total of these two numbers represents the total amount of earth to be moved; their difference represents the excess of cut or filled material.

Retaining Walls

One common purpose of grading is to create flat, usable areas on which to place something. In some cases, where earth banks are as steep as permissible and the amount of available land is limited, one may be forced to construct a retaining wall. A retaining wall creates a level area by cutting vertically through a bank and eliminating the slope necessary to accommodate a soil's angle of repose. Retaining walls are generally constructed of masonry or concrete, although other materials, such as rock, timber, or steel may also be used. A retaining wall is generally an expensive way to create a level area, but where land is costly, the expense may be worthwhile.

When shown in plan, the contours that are cut by a retaining wall appear to stop at one end and emerge at the other. In reality, however, the contours travel along the wall and overlap one another to form a vertical plane of earth that is retained by the wall itself. Spot elevations are usually used to indicate the elevation at the top of wall (T.W.) and bottom of wall (B.W.).

Wing walls are frequently used to accommodate the sloping bank as it returns to the original grade at the ends of the wall. Without wing walls, the bank would wrap around the wall and

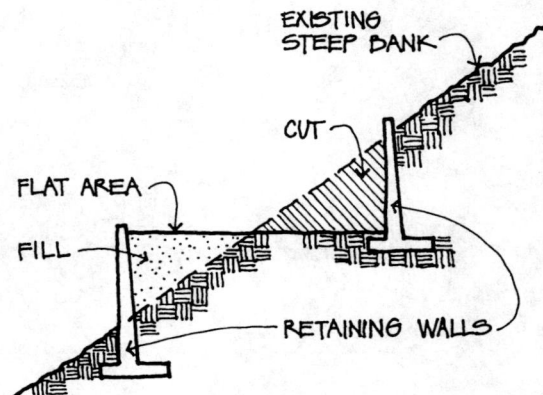

CREATING A FLAT AREA

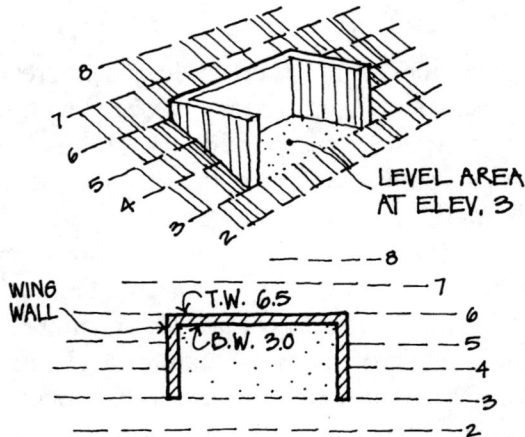

RETAINING WALL

slope down to the original grade. In that case, the angle made by the soil along the face of the wall should not exceed the soil's angle of repose.

Roads

For safety and efficiency, roads are constructed as level as possible; nevertheless, as with pedestrian paths, roads must be slightly sloped to insure adequate drainage. Road drainage is accomplished in the following ways:

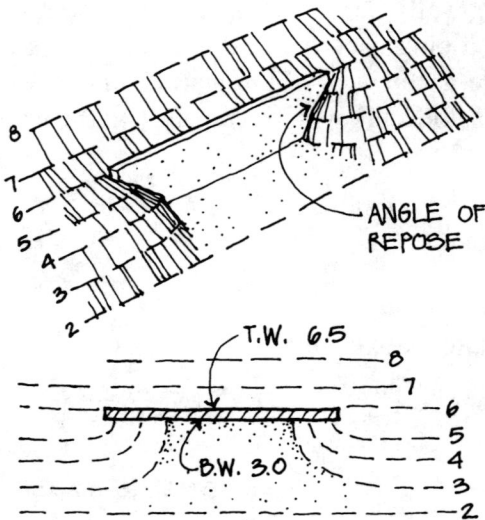

RETAINING WALL

Sloped. The road is graded level across the width but sloped over the length. This causes water to flow in a sheet down the length, where at some point it is collected and conducted elsewhere. Contours are shown crossing the road perpendicularly and break in the downhill direction on each side where the curb is raised.

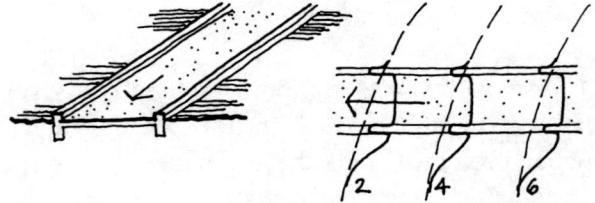

SLOPED

Pitched. The road is pitched toward one side allowing water to flow along one gutter. This method is advantageous where the road parallels the contours, since some of the grade may be taken up by tilting the road in the same direction as the land slopes. Contours cross the road at an angle, and the water flows perpendicular to the contours.

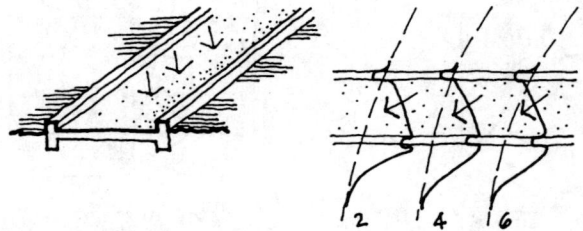

PITCHED

Central Gutter. The road is depressed in the center, which allows the water to run to the middle and flow to some collection area. Contours are drawn similar to a valley, that is, with the contours pointing in the uphill direction.

Central Crown. The road is a convex curve with water pitched toward both sides. This is the most common method used, and a six-inch high

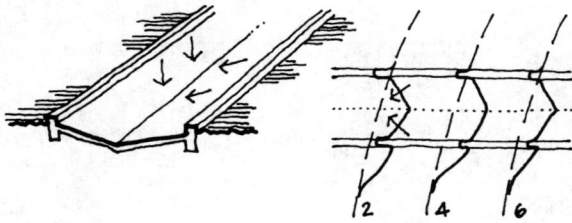

CENTRAL GUTTER

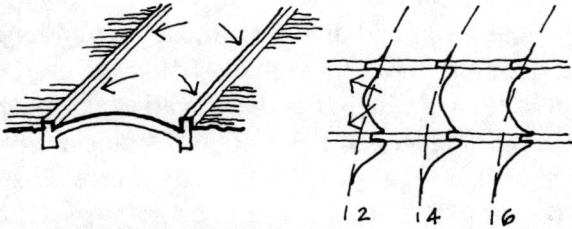

CENTRAL CROWN

crown is typical in a 40 foot street width. Contours are curved and point downhill, similar to the indication for a ridge.

Site designers often have difficulty visualizing contours along a road. Shown below is a contour running across a road with a crown and a raised curb. The contour line curves in the downhill direction, runs level along the curb face, and reappears on top of the curb a little further downhill. Remember, every point along the contour line is at the same elevation.

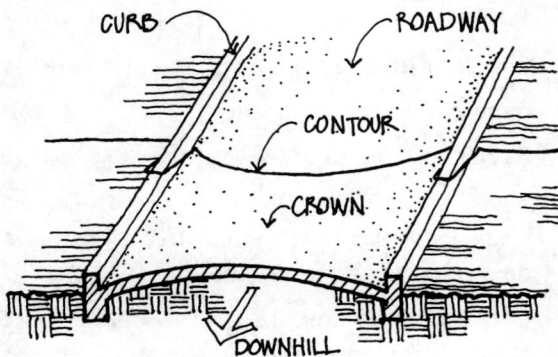

ROAD WITH RAISED CURB

Another example shows a road with a crown and the roadbed elevated above the adjacent grade. The contour line again curves downhill, but reappears on grade a short distance uphill. Other such configurations can be analyzed and envisioned similarly.

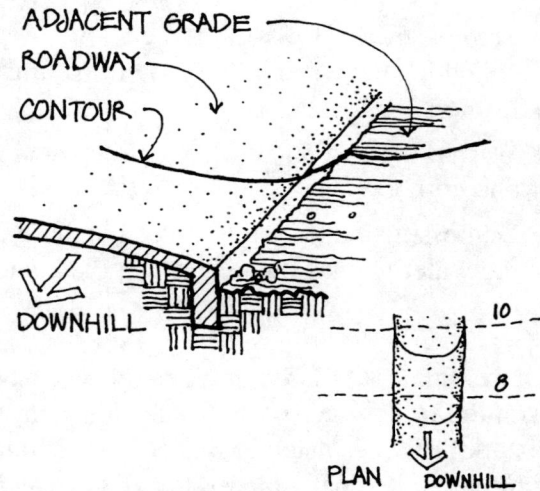

ELEVATED ROAD

Parking Lots

One of the principal objectives of grading is to provide satisfactory surface drainage so that water will be collected and conducted away from the site. For a relatively flat area, such as a parking lot, there are several acceptable ways to do this. First of all, large paved areas should have a minimum grade of 1 percent and a maximum grade of 5 percent. Less than 1 percent will result in occasional ponding, while a slope exceeding 5 percent may allow surface water to flow fast enough to cause erosion or damage.

Parking lots are drained in the following ways:

1. Tipping the entire lot and allowing the water to flow (sheet) along the entire lot length.
2. Tilting the lot and draining the water to one side.

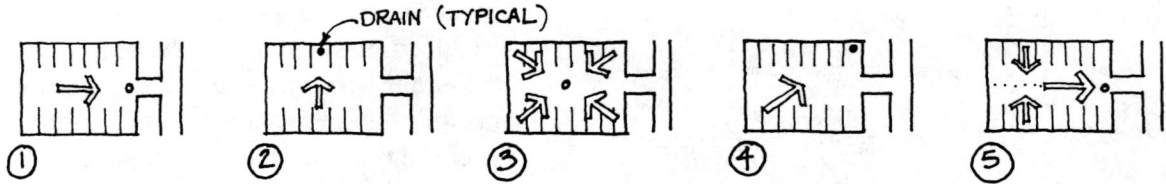

DRAIN (TYPICAL)

① ② ③ ④ ⑤

PARKING LOT DRAINAGE

3. Sloping the lot from all corners and collecting the water at the center in a funnel fashion.
4. Warping the lot and collecting the water in one corner.
5. Depressing the lot at the center and draining the water toward one end along the center line.

In no event should large amounts of water be permitted to flow across sidewalks adjacent to the parking areas. Catch basins or trench drains should be strategically placed to collect flowing water and conduct the excess runoff through the storm drainage system.

Building and Topography

It is always easier and more economical to locate a building on a relatively level site than on one that is steeply sloping. On a level site one has flexibility in the arrangement of building forms. As the ground surface becomes steeper, the site becomes more restrictive and development costs increase. The way a building relates to a sloping site depends largely on the steepness of the slope. Generally, the building footprint should run with its long axis parallel to the contours. This arrangement minimizes the amount of grading required to fit the building to the ground.

The ground directly adjacent to the base of a building should slope away for a short distance so that surface water flows clear of the structure.

This may necessitate creating a swale on the uphill side of the building in order to catch the flowing water and divert it around the building. In addition, the ground floor should always be set at least six inches above the adjacent finish grade to reduce the possibility of water inflow. The situation at access points, of course, is an exception; the finish grade at doorways should be at approximately the same level as the ground floor to allow easy access for the handicapped.

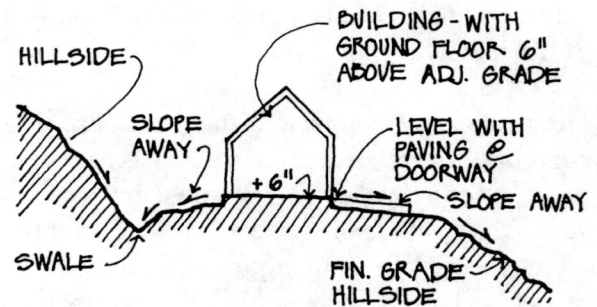

HILLSIDE

SLOPE AWAY

SWALE

BUILDING - WITH GROUND FLOOR 6" ABOVE ADJ. GRADE

LEVEL WITH PAVING @ DOORWAY

+6"

SLOPE AWAY

FIN. GRADE HILLSIDE

SURFACE WATER CONSIDERATIONS

There are several ways to adapt a building to a slope, such as:

1. Create a level area by cutting and filling.
2. Create a level area by using retaining walls.
3. Create two level areas and use a split level design.
4. Create two level areas and use a two-story design over one of them.
5. Create a level area, project the building beyond this area, and elevate it above the ground.

CUT & FILL

RETAINING WALLS

SPLIT LEVEL

TWO STORY

ELEVATED

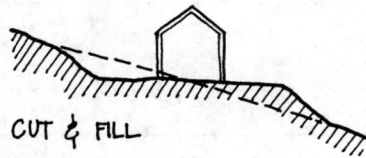

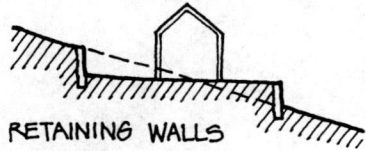

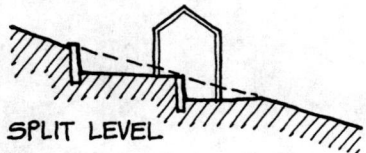

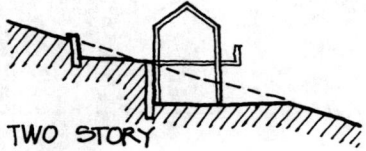

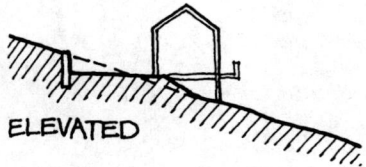

BUILDING ON A SLOPE

In order to achieve a totally successful design solution, buildings must relate to their site both functionally and aesthetically. Designers, therefore, must use their technical knowledge, as well as professional judgment, in dealing with the problems and opportunities created by topography.

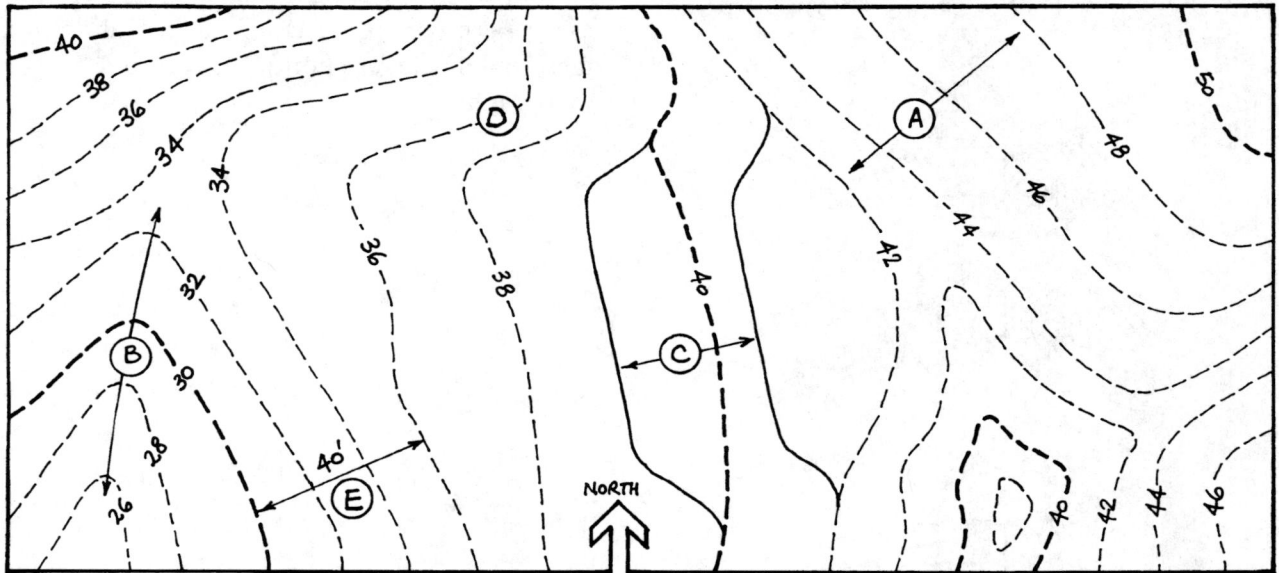

Questions 1 through 5 refer to the topographic map above.

1. The configuration at A is a
   A. concave slope.
   B. convex slope.
   C. uniform slope.
   D. valley.

2. The configuration at B is a
   A. valley.   C. summit.
   B. ridge.   D. depression.

3. The proposed contours at C indicate
   A. about two feet of cut.
   B. about two feet of fill.
   C. about one foot of cut.
   D. about one foot of fill.

4. The drainage at D flows in which direction?
   A. South
   B. Southeast
   C. Northwest
   D. Northeast

5. The gradient at E is approximately
   A. 15 percent.   C. 6.7 percent.
   B. 15°.   D. 6.7°.

6. Which of the following statements are correct?

   I.   Cut earth is generally more stable than filled earth.

   II. Filled earth is generally more stable than cut earth.

   III. Cut slopes are generally permitted to be steeper than fill slopes.

   IV. Fill slopes are generally permitted to be steeper than cut slopes.

   A. I and III   **C.** II and III

   B. I and IV   **D.** II and IV

7. Creating a level area always results in a cut or fill bank which

   A. has the same slope as the existing grade.

   B. is shallower than the existing grade.

   C. is steeper than the existing grade.

   D. may be steeper or shallower than the existing grade.

8. Spot elevations are generally used to indicate

   I.   invert elevations.

   II. tops of curbs.

   III. slope gradients.

   IV. building corners at grade.

   V.   paved ramp surfaces.

   A. I and II   **C.** I, II, and IV

   B. II, IV, and V   **D.** II, III, IV, and V

9. Buildings are often sited parallel to the contours because that orientation usually

   A. maximizes the available level area for construction.

   B. maximizes the solar advantage in most latitudes.

   C. minimizes the amount of disturbance to the natural surface drainage.

   D. minimizes the amount of grading required to fit building to site.

10. Storm drainage systems are designed to achieve all of the following, EXCEPT

   A. reduce erosion.

   B. improve soil texture.

   C. enhance plant growth.

   D. eliminate standing water.

CIRCULATION

MOVEMENT SYSTEMS

Introduction

Circulation systems are paths of movement that connect the activities and uses of an area. The improvement of a site depends largely on its accessibility to people, vehicles, and the utilities necessary to serve it. Without access, land has little practical value, no matter how beautiful or abundant in resources it may be. Circulation systems allow the elements of a site to function properly. These systems must be related to natural or man-made topography: they must take into account the nature of the terrain and the natural features or structures that are part of it.

The design of each system must relate to the tempo of the movement it accommodates, as well as to the general nature of its surroundings. Vehicular systems require free-flowing forms in response to the rhythm of fast movement. Pedestrian systems require interest, variety, and impressions of rapid change, through the use of focal points, visual termini, and so on. Furthermore, the layout of circulation systems is dictated by the needs of its users, who must be moved efficiently to satisfy the functional demands of the program.

The flexibility of a circulation system affects the future development and growth patterns of an area, both within and beyond the limits of the site. Finally, site circulation systems must be in harmony with the surroundings to allow a safe and smooth flow of traffic and services.

Vehicular Circulation Patterns

Circulation patterns are influenced by the concentration and density of population, the topography, and the built environment.

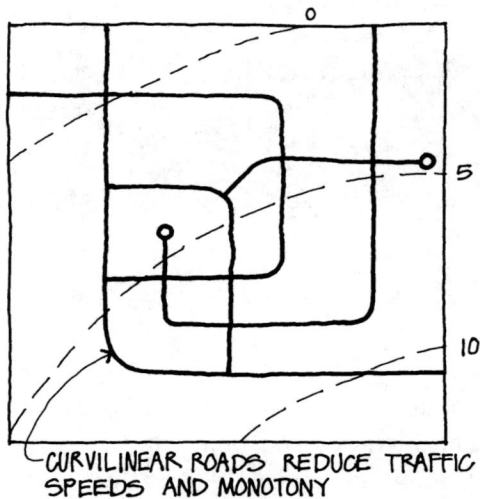

CURVILINEAR ROADS REDUCE TRAFFIC SPEEDS AND MONOTONY

REGULAR TOPOGRAPHY

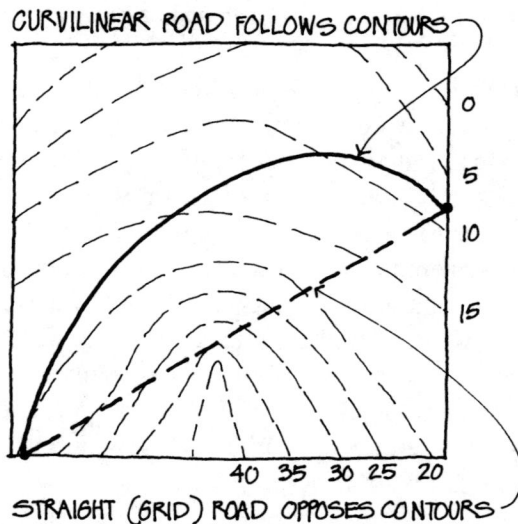

CURVILINEAR ROAD FOLLOWS CONTOURS

STRAIGHT (GRID) ROAD OPPOSES CONTOURS

IRREGULAR TOPOGRAPHY

Sometimes, these patterns ignore logic and simply follow historical models. In San Francisco, for example, a grid pattern is superimposed on the relatively steep and irregular topography of the city. Anyone who has climbed one of the city's streets knows the problems caused by a design concept that ignores the form of the land. A *grid* system usually consists of equally spaced streets, perpendicular

to each other. Its geometry is regular and simple, and orientation and direction are easy to comprehend. On level ground the grid system produces a clear pattern of regularly shaped properties and straight circulation systems. It is, however, less interesting visually than one more sympathetic to the natural topography. Other systems may follow *radial, linear,* or *curvilinear* patterns. Radial and linear patterns are limited because of their rigid geometry. Where the landscape is irregular, curvilinear patterns follow the contours of the land as closely as possible. Curvilinear streets may also be appropriate in relatively flat residential neighborhoods, in order to reduce traffic speeds and monotony.

Access to urban sites is usually from the surrounding grid system comprised of local and collector streets. Suburban and rural sites are accessible from access streets fed by highways, expressways, or freeways.

Existing traffic patterns, streets, directions of flow, relative speeds, and public transportation routes must be considered in evaluating alternative vehicular and pedestrian access points.

One may have to choose the points of access which provide the greatest safety for both pedestrians and vehicles. The choice may be between a high-speed, multi-lane street and another with enforced speed limits resulting in slower traffic movement, carrying one-way traffic only. The latter provides access from one direction only, but may offer greater safety of ingress and egress for vehicles. The ultimate decision may be further complicated by the volume of traffic generated, especially at peak periods in the mornings when schools, factories, and businesses begin to operate and in the afternoons when they let out.

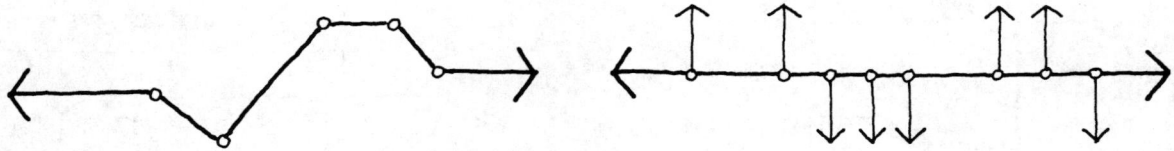

LINEAR PATTERNS

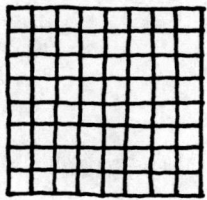

GRID PATTERNS

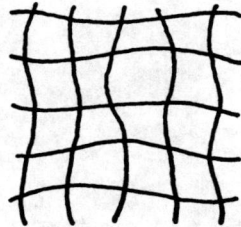

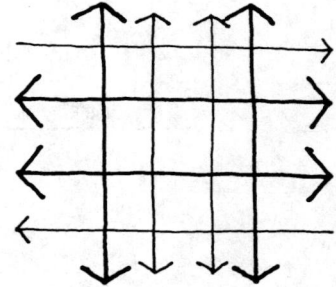

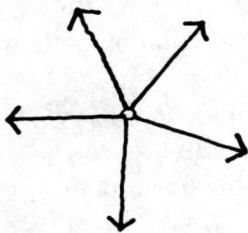

RADIAL PATTERNS

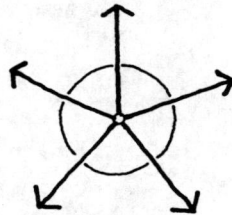

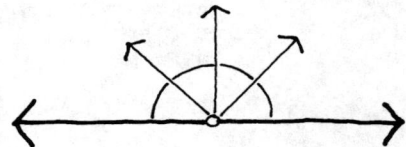

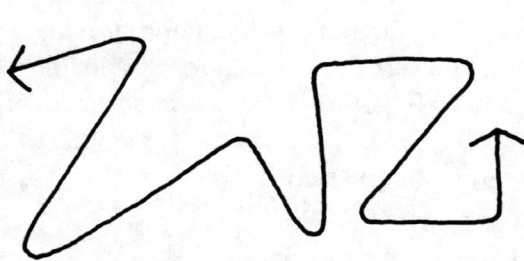

CURVILINEAR PATTERNS

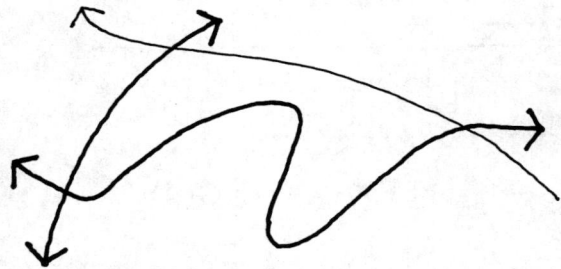

PATTERNS OF CIRCULATION

Relating access points to traffic lights, intersections, signs, and pedestrian crossings is vitally important to public safety. For example, if a new school is constructed, additional traffic controls may be required to allow school children to safely cross existing streets. A shopping center may require a pedestrian bridge between a parking structure and the mall. It is common to

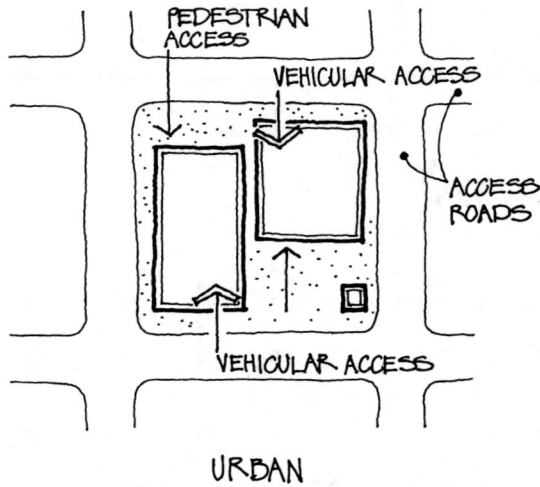

PEDESTRIAN
ACCESS

VEHICULAR ACCESS

ACCESS
ROADS

VEHICULAR ACCESS

URBAN

ACCESS ROADS

HIGHWAY

P   P

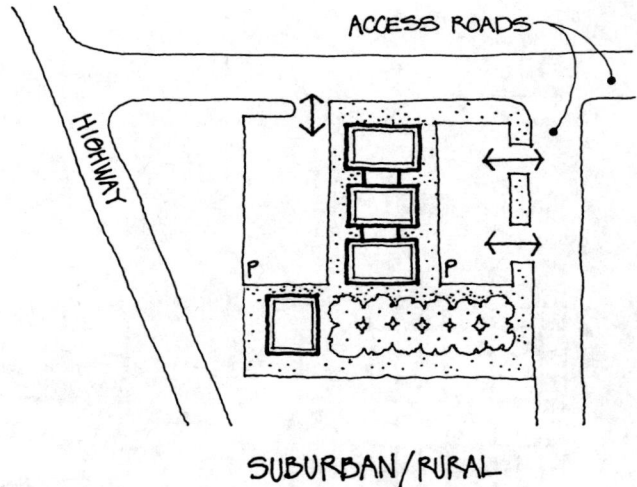

SUBURBAN/RURAL

SITE ACCESS

NEW TRAFFIC SIGNAL
FOR PEDESTRIAN
CROSSING

PEDESTRIANS   HIGHWAY

ACCESS
STREET   ACCESS

SCHOOL SITE

NEW TRAFFIC
SIGNAL FOR
PEDESTRIAN
CROSSING

PEDESTRIANS

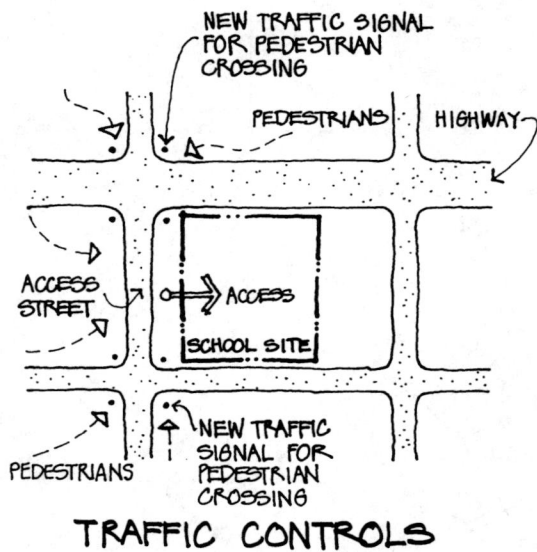

TRAFFIC CONTROLS

vertically separate arriving and departing vehicles at airports and to provide connecting bridges or underpasses for pedestrians circulating to and from parking structures.

Vehicular on-site circulation systems must provide easy and direct access for both service and passenger vehicles, which should be separated

whenever possible to avoid cross traffic and potential hazards. The circulation pattern on a site should respond to the specific needs of the site development and may be quite different from that of the surrounding streets. The designer has considerable freedom in laying out the site circulation paths, so long as they join the surrounding system at the site's periphery. Basic design criteria for slopes, widths, alignment, and drainage must, nevertheless, be maintained to facilitate movement and provide safety. The vehicular circulation system must be suitable to fulfill its basic task, appear orderly and well oriented, and connect smoothly with its surrounding systems.

Pedestrian Circulation Patterns

The movement of pedestrians plays an important role in locating structures and organizing improvements on a site. On a university campus, for example, students and faculty must be able to walk between buildings and other campus facilities. To develop safe, comfortable, and interesting circulation paths, it is necessary to simultaneously consider function, topography,

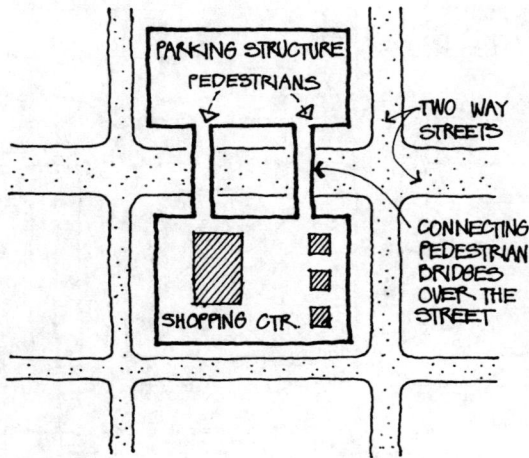

OVERPASS

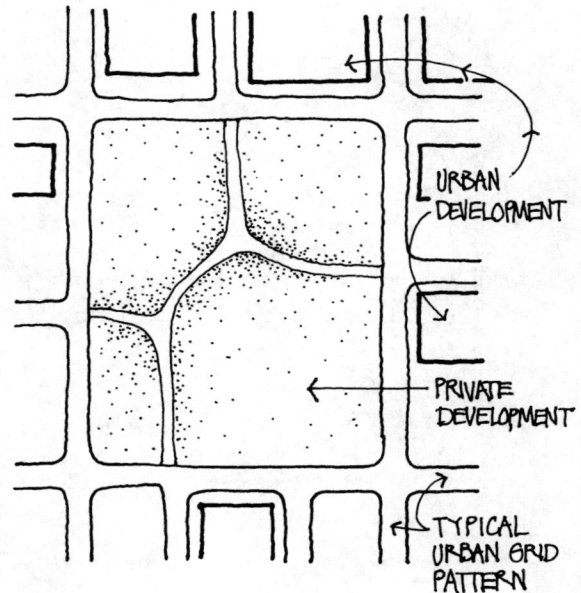

INTERFACE OF VEHICULAR
CIRCULATION SYSTEMS

climate, and visual perception. Furthermore, one should anticipate the preferred routes so that pedestrians will not beat a path through lawns rather than use the paved walks. These paths must provide a clear visual definition to prevent pedestrians from getting lost—signs can never compensate for a confusing network of walks.

Safety is a product of good design— sensibly-sized walkways, properly sloped and appropriately finished with durable non-slip

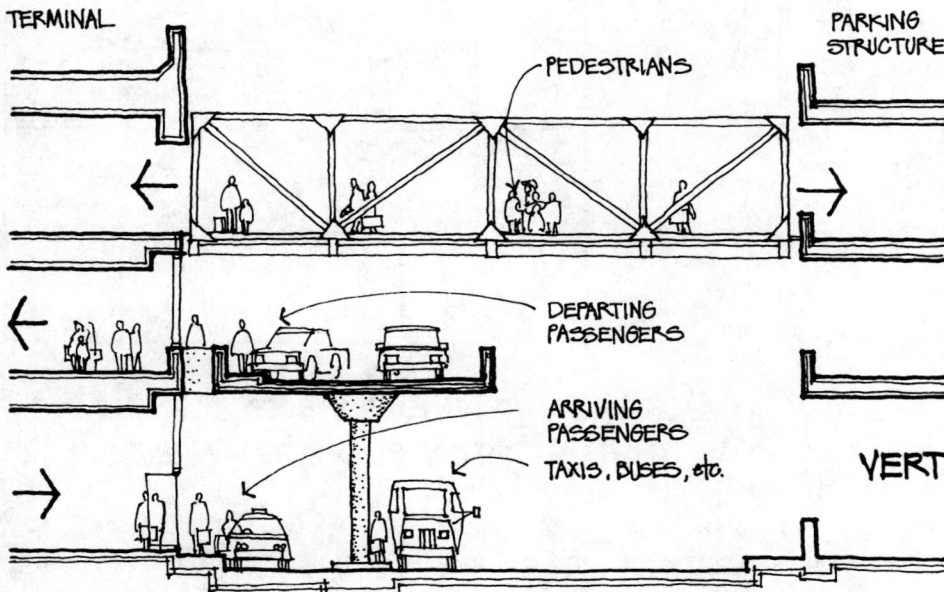

VERTICAL SEPARATION

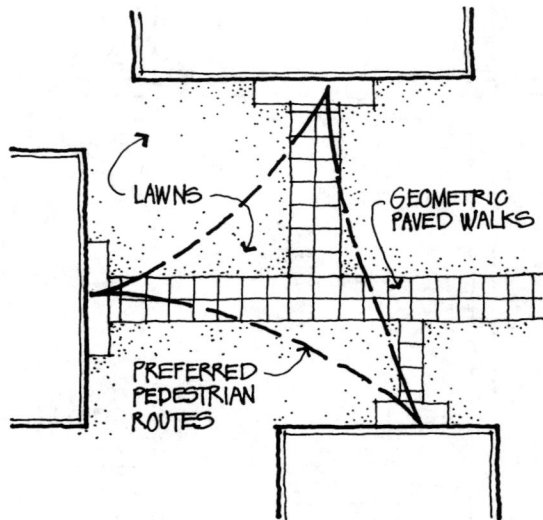

PEDESTRIAN CIRCULATION PATTERNS

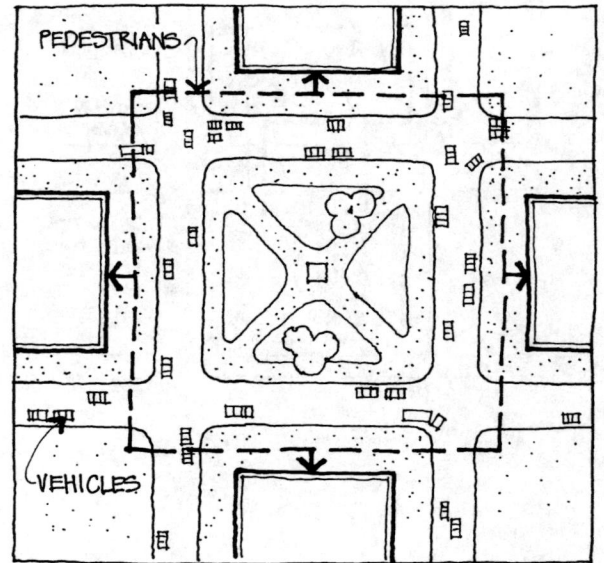

NO VEHICULAR/PEDESTRIAN SEPARATION

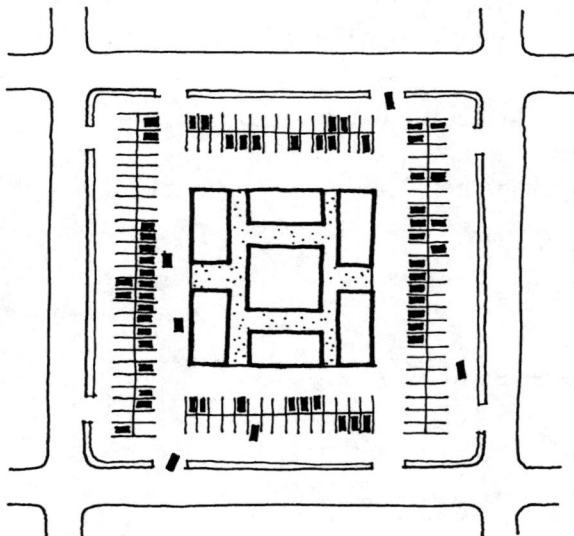

PERIMETER PARKING

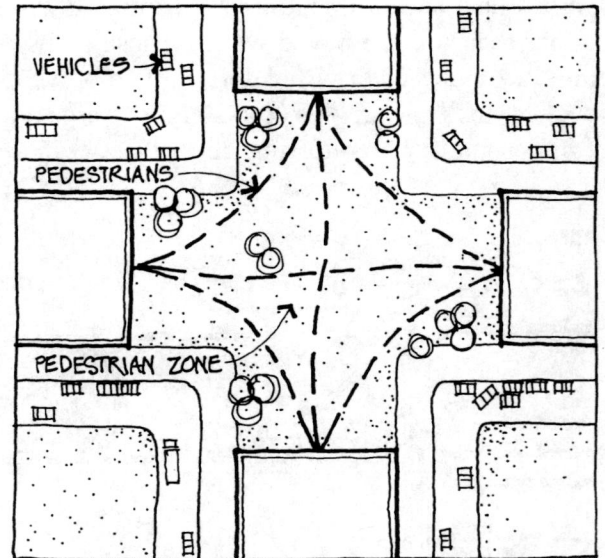

VEHICLES EXCLUDED FROM PEDESTRIAN ZONE

materials. Security and convenience result from clear sight lines, good lighting, and the placement of benches for resting along the way. Planting, decorative paving, pools, fountains, and site furniture enhance the aesthetic experience of circulating between site elements. Walkways may lead to or cross malls, courts, or green spaces which serve special outdoor uses. These spaces must be designed to act as connecting circulation elements, as well as places for assembly, performance, exhibit, display, and similar functions. For example, the paved areas in a sculpture garden must allow pedestrians to view the works of art without interfering with the normal flow of foot traffic. Such spaces also relate visually to adjacent buildings; consequently they must be in proportion to their surroundings in order to provide the proper spatial definition and feeling of enclosure.

For some types of development, the pedestrian circulation system becomes the dominant structuring element. This holds true where pedestrians have the highest priority and vehicles are prevented from intruding into their zone. Regardless of priority, a site design for a complex of related uses must avoid conflicts between pedestrians and vehicles, provide an efficient link between its various facilities, and reinforce the organization of its individual elements.

CIRCULATION SYSTEMS

Introduction

Paths of movement, whether of people, automobiles, goods, or services, are linear in nature. They all have a starting point from which they move through and past a sequence of spaces until they arrive at a destination. The shape and form of the path depend on the type of transportation. Pedestrians can turn, pause, stop, and rest at will. An automobile, however, has less freedom to change direction or come to a sudden halt. Nevertheless, an automobile can negotiate a smoothly contoured path tailored to its physical size.

Pedestrians, on the other hand, although able to tolerate abrupt directional changes, require more space than their physical size and greater freedom to choose their direction of movement. Vehicular traffic can also be more easily controlled than foot traffic, since automobiles

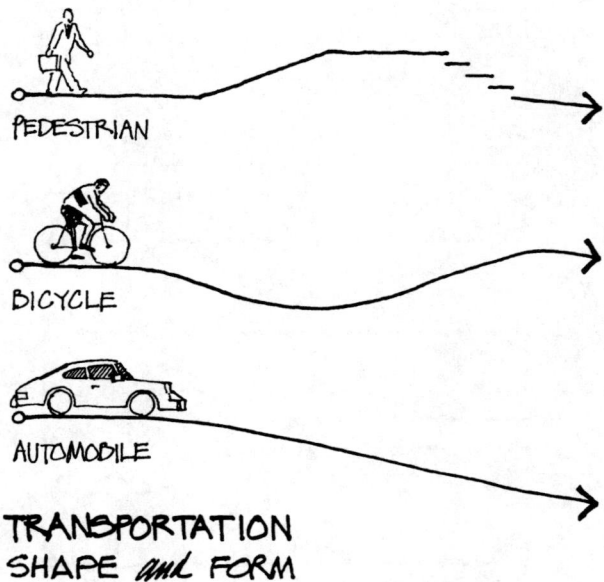

PEDESTRIAN

BICYCLE

AUTOMOBILE

TRANSPORTATION SHAPE and FORM

do not have as many choices of circulation routes nor the freedom of access that pedestrians have.

When one reaches a branch or an intersection in a walkway or road, one must decide in which direction to proceed. The scale and continuity of each path at an intersection allow one to distinguish between major and minor routes leading to more or less important buildings and spaces. When intersecting roads and walks are similar in size, people must pause, orient themselves, and decide which direction to follow, and the designer must provide sufficient space for this. The configuration of roads and walks relates directly to the pattern of buildings and spaces it links. A circulation element that parallels a pattern of buildings reinforces the spatial organization; one that opposes it can act as a visual counterpoint to such a pattern. If one is able to visualize a site's overall layout and configuration of roads and walks, the orientation and spatial arrangement are clearly perceived. Consequently, one of the primary objectives of a circulation system is to lead rather than to confuse.

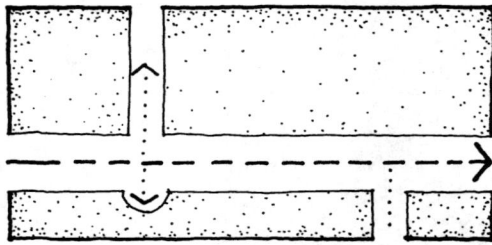

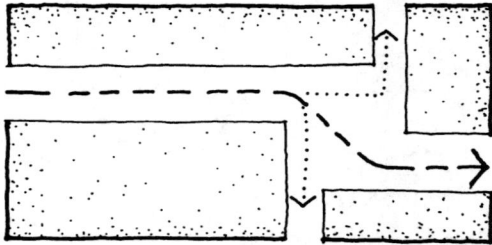

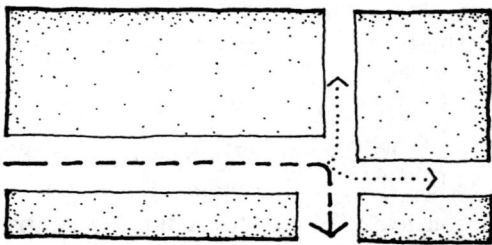

BRANCHES and INTERSECTIONS

Design of Vehicular Circulation

Virtually all developable sites are near or adjacent to public streets. Most urban sites are accessible from streets along one property line, which limits the number of choices for locating points of ingress and egress. Suburban and rural sites usually offer more choices, although they may require the construction of special access roads to connect with existing highways. In every case, the following are primary considerations in locating points of access to a site:

1. Interface points of access with surrounding circulation systems and patterns.
2. Relate access points to existing and future adjacent uses and developments.
3. Avoid potential conflicts between vehicles and pedestrians.
4. Relate access points to on-site parking and service areas.
5. Minimize the environmental impact of points of access on natural features of the site and its surroundings.

The architect must try to solve all of these access problems, or at least provide the best possible compromise solution. On the issue of public safety, however, there are no compromises.

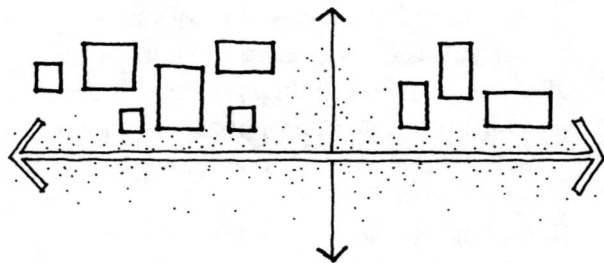

PARALLELS

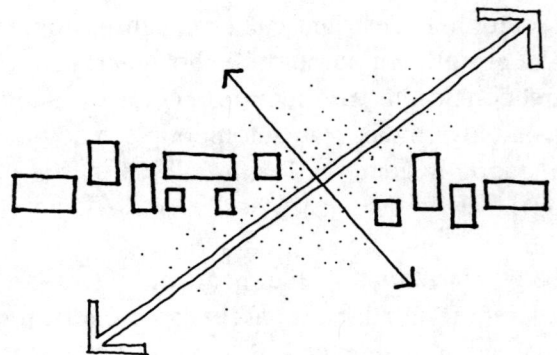

OPPOSES

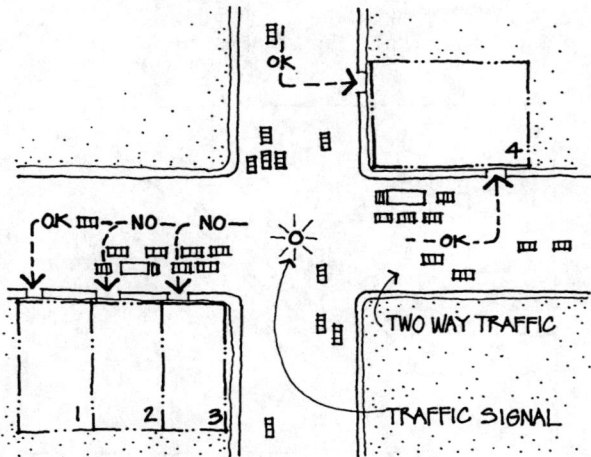

LEFT TURN SITE ACCESS

One of the objectives in locating vehicular access points in urban areas is to allow cars to turn either left or right when entering or leaving the site. This is most critical where the amount of street frontage is limited and the site is near a major street intersection.

A sufficient distance between the driveway and the intersection will permit left turn access onto the site even when cars are stacked and waiting for a signal change at the intersection. The situation improves as this distance increases.

The most undesirable access locations are those close to major intersections, since driveways interfere with traffic moving through these intersections. Cars waiting for a signal change block the driveway entrance when the driveway is less than 200 feet from the intersection. Left turn access is also blocked during peak periods when waiting traffic backs up past the driveway. Furthermore, vehicles unable to enter the driveway block the flow of traffic in two directions, causing a virtual standstill. And pedestrians using sidewalks near intersections are exposed to greater potential danger. Access points to sites on opposite sides of streets should avoid interfering with each other. It is preferable to locate

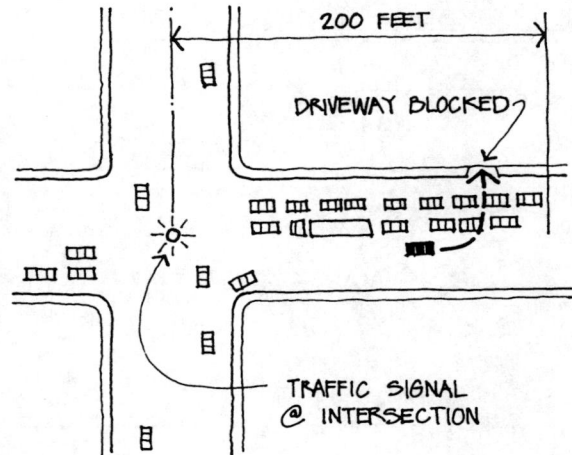

MINIMUM DISTANCE TO DRIVEWAYS

driveways directly opposite each other where both share a single access location, similar to a street intersection. Where such an arrangement is not feasible, access points should be offset to provide adequate left turn stacking space in advance of each driveway and to avoid the overlap of left turn lanes.

The location of driveways must also consider adjacent developments. For example, if the site is next to an elementary school where children use sidewalks and street crossings, driveways should be located to avoid circulation conflicts, in order to maximize pedestrian safety.

Suburban and rural sites may offer a wider choice for the location of access points. Here, the main consideration might be the preservation of the natural environment. If given a choice, for example, one would not destroy a mature grove of trees to cut a path for roads; one would, more likely, select a route that passes over less valuable land. Furthermore, grading and paving of roads is costly, consequently, it is always best to locate driveways and align roads in harmony with existing contours. If a great deal of grading is needed because the road opposes the natural topography of the site, the

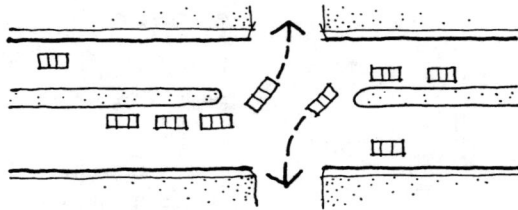

DRIVEWAYS OPPOSITE

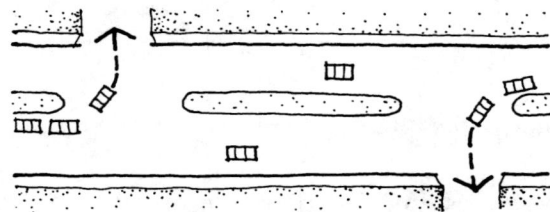

APPROPRIATE

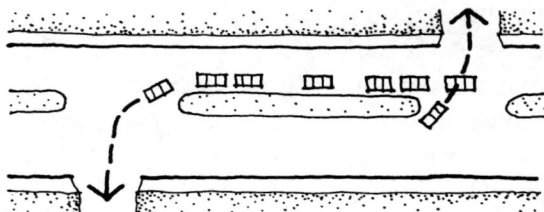

INAPPROPRIATE

DRIVEWAYS STAGGERED

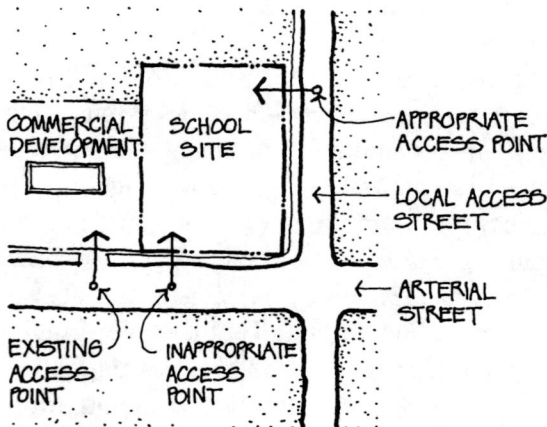

COMMERCIAL DEVELOPMENT

SCHOOL SITE

APPROPRIATE ACCESS POINT

LOCAL ACCESS STREET

ARTERIAL STREET

EXISTING ACCESS POINT

INAPPROPRIATE ACCESS POINT

ACCESS - ADJACENT USES

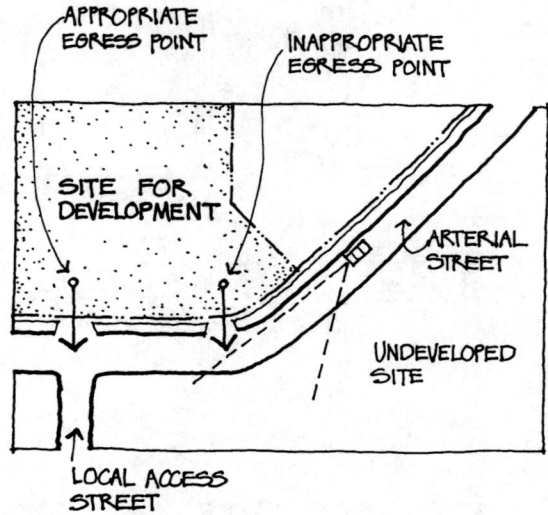

APPROPRIATE EGRESS POINT

INAPPROPRIATE EGRESS POINT

SITE FOR DEVELOPMENT

ARTERIAL STREET

UNDEVELOPED SITE

LOCAL ACCESS STREET

ACCESS - VISIBILITY

cost will increase, and cutting the natural grade will scar the site's surface and require the construction of berms and banks.

Public safety may be the determining factor in selecting an access point from a highway onto a rural site. The road's alignment and the ability of high-speed traffic to clearly see cars leaving and entering the site may be of greater concern than the convenience of drivers.

When a site is in the vicinity of an expressway interchange, it is preferable to locate access drives to commercial developments along arterial and collector streets at a considerable distance from the interchange. This forces vehicles to reduce speed and allows drivers to become oriented to local traffic patterns and developments. Driveways to residential sites should be from local access streets to further reduce the dangers from traffic.

If, after considering all these factors, there are several suitable access points for vehicles, how does one decide which of these to use? The primary concern should be how these access

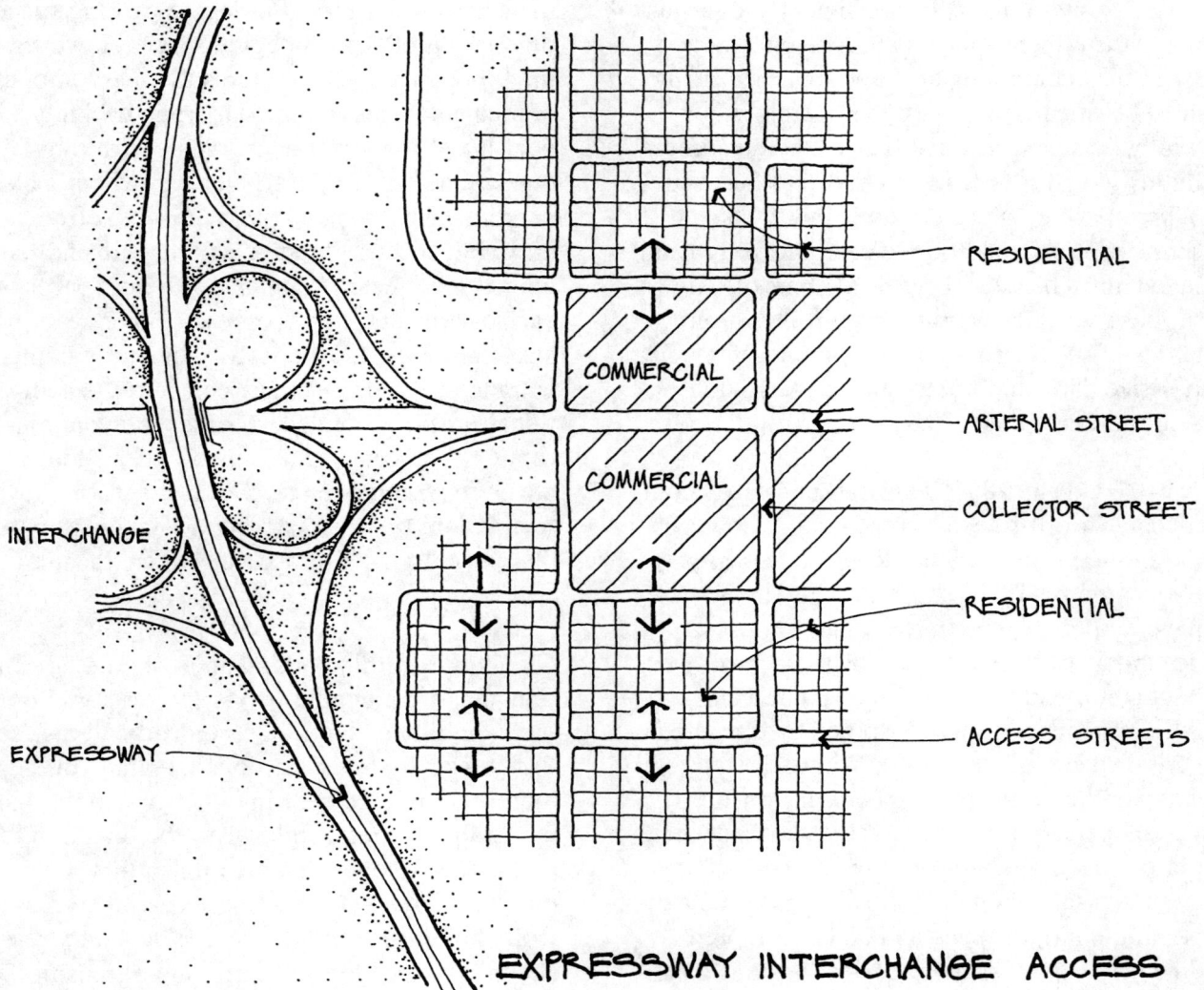

EXPRESSWAY INTERCHANGE ACCESS

points relate to the on-site circulation for passenger and service vehicles and to the required amount and optimum location of parking, and how they interface with each other and the buildings on the site.

Parking

Virtually every development requires automobiles to be parked on its site. Most municipalities have ordinances requiring a developer to provide parking based on usable floor area—for example, two parking spaces for every 1,000 square feet of leasable office area. In some cases, the requirements may be stated in terms of parking spaces per unit use, such as two spaces per apartment. Therefore, parking has a major impact on the functional and aesthetic qualities of the site. Whatever the specific requirement may be, parking areas must be designed to minimize ground coverage, provide safe access from public streets, afford ease of circulation for its users, and be functionally related to the buildings they serve.

To move and store vehicles efficiently, one must apply the dimensional and movement characteristics of automobiles to the design of parking areas. Critical factors are the overall length, width, front and rear overhangs, and the turning radius traced by both front and rear bumpers. Where parking areas are used by trucks and semi-trailers, in addition to cars, the design and layout must be sized to accommodate the largest of these vehicles. Turning radii range from about 16'-0" for sub-compact cars to 25'-0" for full-size cars, and 45' to 50' for trucks and semitrailers.

A 9'-0" wide by 20'-0" long parking stall accommodates a full-size car comfortably, however, these dimensions should never be less than 8'-0" wide by 18'-0" long. Parking areas may have a number of different configurations, depending on their capacity, distribution of cars, circulation pattern, and site characteristics. Aisle dimensions are determined by the maneuvering space required to park in each stall and whether they serve one- or two-way traffic. The width of aisles and parking bays varies, depending on the angle between car and curb. The diagrams shown on the following page indicate parking configurations and aisles, required dimensions, and space needs for various conditions of angled parking.

Parking stalls on either side of an aisle may be at 90 degrees or angled in opposite directions. The latter results in a herringbone pattern and is often used where space limitations prevent perpendicular parking.

In estimating the total area required for parking, it is necessary to include stalls, aisles, and connecting drives. The data shown on the following page are used to compare different parking arrangements and determine a safe and efficient layout. For example, it is more difficult to maneuver a car into a 90-degree parking space than into a 60- or 45-degree space. However, the 90-degree layout requires the least amount of area and is less dangerous when backing out because of the wider aisle. Angled parking layouts create one-way traffic aisles, which make parking easier. Angled parking is less efficient, however, because it requires more curb and stall length and creates triangular leftover spaces at each car and at the ends of rows. Where parking aisles are used for drop-off points at building entrances, the layout must consider the additional traffic generated, in order to avoid disrupting the normal traffic flow. Cars should never be required to use public streets or alleys to circulate between parking aisles, nor should they have to back into streets when vacating spaces.

Circulation within parking areas should be continuous. There should be the fewest possible turns, and there should be no dead-end aisles. Parking areas should be sloped to drain, but not more than five percent in either direction, that is, parallel or perpendicular to the direction of parked cars. Slopes should be uniform whenever possible. Driveways leading to parking areas from streets, or ramps connecting separate areas, should also be sloped, to provide a transition between different elevations. Such ramps should not exceed a slope of ten percent.

All layouts must, of course, conform to the requirements for the physically handicapped. Specifications for these are published by the American National Standards Institute, publication ANSI 117.1. Although the number of handicapped stalls is not specified, a reasonable number must be provided in each parking area. These spaces and their adjacent passenger loading zones must be located to provide the shortest possible circulation route to the building entrance.

PARKING DIMENSIONS AND EFFICIENCY RELATIONSHIPS

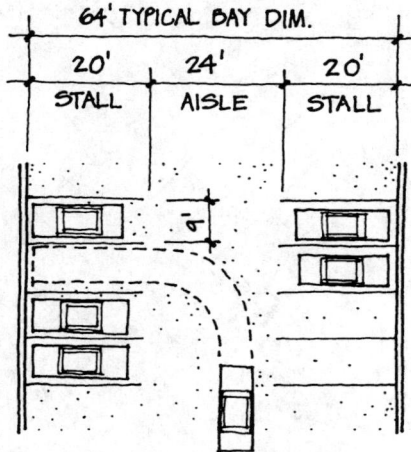

64' TYPICAL BAY DIM.

| 20' STALL | 24' AISLE | 20' STALL |

9'

90° ANGLE (20' × 9' STALLS)

- 11.1 CARS FOR EACH 100 LINEAL FT. OF CURB
- 290 SQ. FT. PER CAR AREA REQUIREMENT
- ACCOMMODATES MOST CARS
- PERMITS 2-WAY TRAFFIC AISLES
- MORE DIFFICULT TO MANEUVER

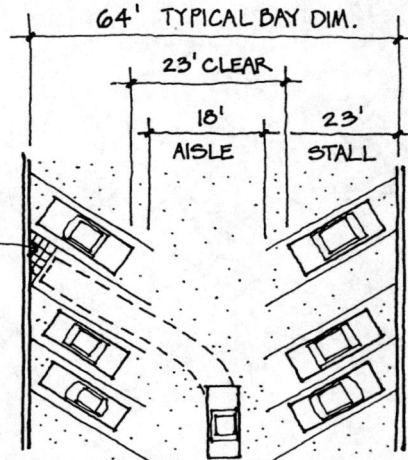

64' TYPICAL BAY DIM.

23' CLEAR

| 18' AISLE | 23' STALL |

TYPICAL WASTED SPACE

60° ANGLE (20' × 9' STALLS)

- 9.7 CARS FOR EACH 100 LINEAL FT. OF CURB
- 333 SQ. FT. PER CAR AREA REQUIREMENT
- EASY ACCESS
- ONE·WAY TRAFFIC AISLES
- MOST POPULAR CONFIGURATION
- RELATIVELY ECONOMICAL

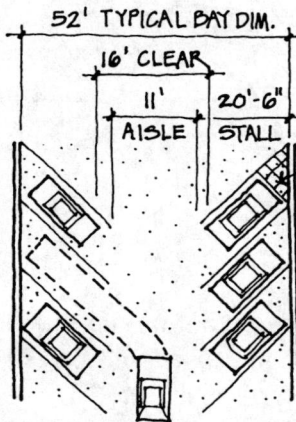

52' TYPICAL BAY DIM.

16' CLEAR

| 11' AISLE | 20'-6" STALL |

TYPICAL WASTED SPACE

45° ANGLE (20' × 9' STALLS)

- 7.8 CARS FOR EACH 100 LINEAL FT. OF CURB
- 333 SQ. FT. PER CAR AREA REQUIREMENT
- EASY ACCESS
- ONE·WAY TRAFFIC AISLES
- RELATIVELY ECONOMICAL

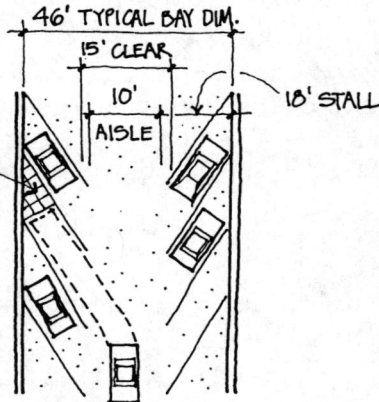

46' TYPICAL BAY DIM.

15' CLEAR

| 10' AISLE | 18' STALL |

30° ANGLE (20' × 9' STALLS)

- 5.5 CARS FOR EACH 100 LINEAL FT. OF CURB
- 414 SQ. FT. PER CAR AREA REQUIREMENT
- EASY ACCESS
- ONE·WAY TRAFFIC AISLES
- RELATIVELY UNECONOMICAL

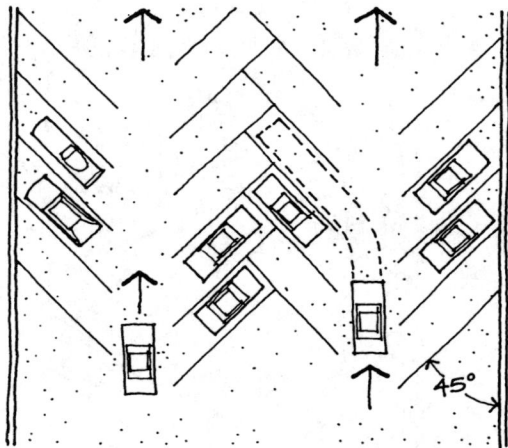

ONE·WAY AISLES ; SAME DIRECTION
OF TRAVEL IN EACH AISLE.

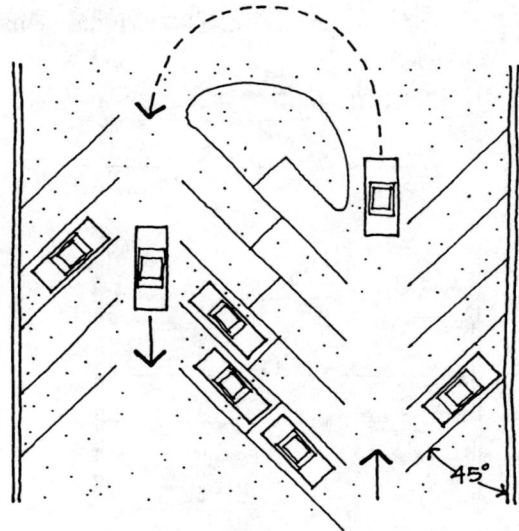

ONE·WAY AISLES ; OPPOSITE DIRECTION
OF TRAVEL IN ALTERNATE AISLES.

HERRINGBONE PARKING PATTERNS

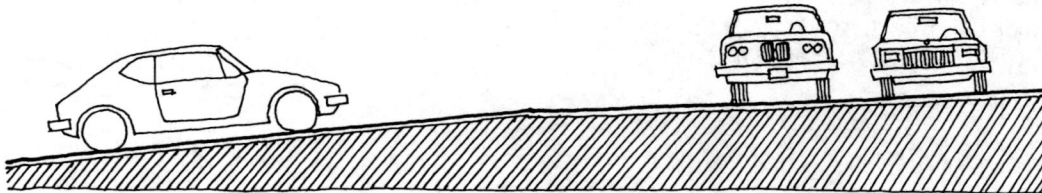

AUTO DRIVEWAY RAMP SLOPE :
I FOOT RISE IN 10' RUN
OR 1:10 (10% SLOPE)

PARKING LOT SLOPE :
I FOOT RISE IN 20' RUN
OR 1:20 (5% SLOPE)

MAXIMUM SLOPES FOR VEHICLES

The minimum dimensions of handicapped parking stalls are 20 feet long and 8 feet wide, with a passenger loading zone no less than 5 feet wide adjacent and parallel to the vehicle pull-up space, for a total width of 13 feet. The five-foot wide loading zone may be located between two handicapped stalls to serve both, resulting in a total width of 21 feet for the two 8-foot stalls and the 5-foot loading zone. An accessible route to the building must be located at the front of the stall, to avoid the hazard of handicapped persons having to circulate behind parked vehicles.

The typical parking layout on the following page shows the handicapped and other criteria for on-site parking of vehicles.

It is preferable to provide separate service drives leading to service points whenever possible. Where service vehicles must share the aisles and drives with passenger vehicles, their routes should be as short as possible to minimize potential conflicts. Loading areas should be out of the way of vehicular routes and provide ample space for service vehicles to reverse and turn.

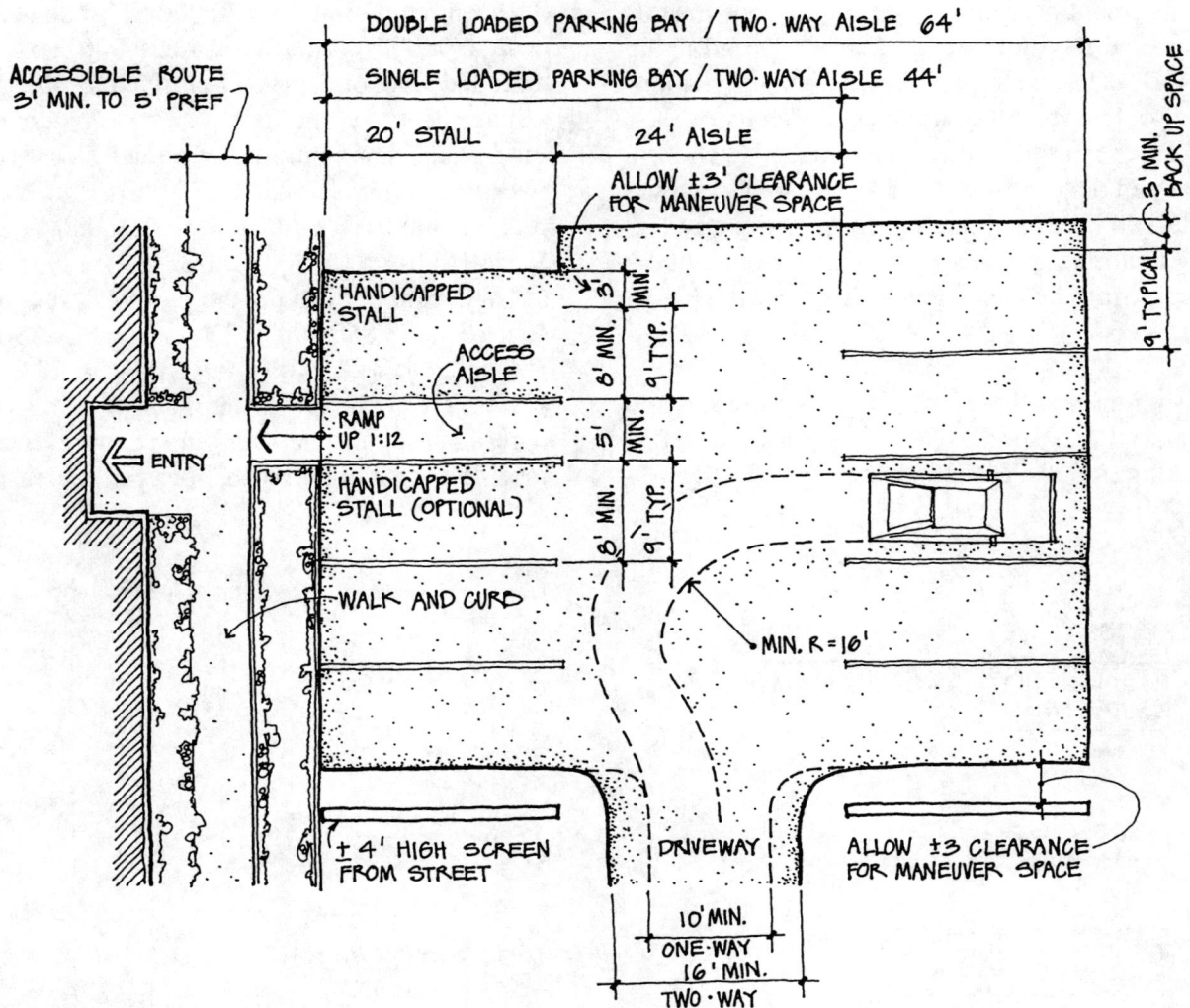

Diagram labels:

DOUBLE LOADED PARKING BAY / TWO·WAY AISLE 64'

SINGLE LOADED PARKING BAY / TWO·WAY AISLE 44'

ACCESSIBLE ROUTE 3' MIN. TO 5' PREF

20' STALL   24' AISLE

ALLOW ±3' CLEARANCE FOR MANEUVER SPACE

3' MIN. BACK UP SPACE

9' TYPICAL

HANDICAPPED STALL

ACCESS AISLE

8' MIN.   3' MIN.   9' TYP.

RAMP UP 1:12

5' MIN.

ENTRY

HANDICAPPED STALL (OPTIONAL)

8' MIN.   9' TYP.

WALK AND CURB

MIN. R = 16'

± 4' HIGH SCREEN FROM STREET

DRIVEWAY

ALLOW ±3 CLEARANCE FOR MANEUVER SPACE

10' MIN. ONE·WAY

16' MIN. TWO·WAY

TYPICAL PARKING LAYOUT (PREFERRED DIMENSIONS)

Passengers must be able to circulate safely between automobiles and buildings. Paths may be defined by striping applied to paving, locating raised walkways between bays, and, in heavy traffic situations, pedestrian bridges.

Parking areas used at night should be uniformly illuminated to an intensity of one-half foot candle by regularly spaced lighting standards, 30 to 50 feet in height. Fixtures should be placed to provide an overlap of light patterns and avoid dark areas. The light source for parking lot lighting is usually mercury vapor or high-pressure sodium, for efficiency and economy. The cost of parking includes the cost to acquire the land, grading, paving, curb, drainage, striping, and lighting.

If land costs exceed the cost to build parking structures, it may be more economical to stack layers of cars than to park all of them on grade. An analysis using the average gross area requirement per vehicle, the square foot cost of land, and the unit cost to improve the ground, as well as to construct parking structures enables one to determine the more cost-effective solution. For example, you might be asked to choose between acquiring sufficient land to accommodate 200 cars on grade or constructing a two-level parking garage on one-half the amount of land, using the cost figures shown in the chart below.

Assuming that the average total area requirement is 350 SF per car, the total area needed for parking is 350 SF × 200 cars = 70,000 SF.

To acquire and improve 70,000 SF of land will cost 70,000 SF × $66/SF = $4,620,000. To build a second level of parking would reduce the land requirements by one-half. The cost to accommodate the same number of vehicles for a two-level scheme would be 35,000 SF × $66/SF plus 35,000 SF × $44/SF = $2,310,000 + $1,540,000 = $3,850,000. Consequently, the average cost per car for the parking structure scheme is $3,850,000 ÷ 200 cars = $19,250, compared to $4,620,000 ÷ 200 cars, or $23,100 per car on grade. As the number of levels of parking structure increases to accommodate more cars, the average cost per car decreases.

AVERAGE
COST / S.F.

ALL ON GRADE

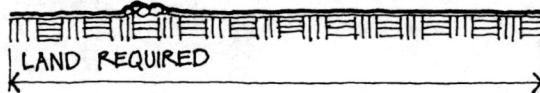

LAND REQUIRED

$54 / S.F.   LAND COST

$12 / S.F.   IMPROVEMENTS

$66

TWO LEVEL

LAND REQUIRED (½)

$44 / S.F.   STRUCTURE

$54 / S.F.   LAND COST

$12 / S.F.   IMPROVEMENTS

$55

FOUR LEVEL

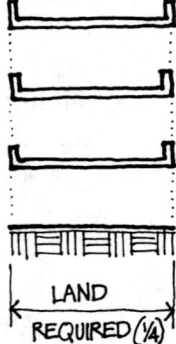

LAND
REQUIRED (¼)

$44 / S.F.   STRUCTURE

"   "

"   "

$54 / S.F.   LAND COST

$12 / SF.   IMPROVEMENTS

$49.50

PARKING COSTS

Design of Pedestrian Circulation

In downtown locations, building entrances are usually accessible to pedestrians from adjacent public sidewalks. The desire to retain more open space in the cities has prompted developers to sacrifice portions of the buildable area of sites for landscaped plazas, pocket parks, fountains, pools, and similar outdoor areas designed to enhance the environment. Consequently, pedestrian spaces in the city are not necessarily limited to sidewalks.

Furthermore, many cities have barred the automobile from certain commercial zones and created landscaped malls for the pleasure and safety of pedestrians, providing opportunities for social interaction and visual enjoyment. The design of these and other pedestrian facilities requires an understanding of the physical dimensions and movement characteristics of the human body under a variety of circumstances.

For example, the area covered by a human being standing still is approximately three square feet. This is based on an assumed shoulder breadth of 24 inches and a body depth of 18 inches for the average adult male. In order to move easily in a crowd without creating body contact, a total of 13 square feet per person is required. A lesser area will normally impede movement and require an effort on the part of pedestrians to avoid contact while walking. If the area allowance is less than seven square feet, pedestrians are forced to move as a group rather than to move as individuals unimpeded in all directions. If the area is only three square feet per person, no movement is possible, body contact is inevitable, and a dangerous situation may develop in case of panic caused by a sudden need to reach an exit.

Human motion involves balance, timing, and sight. Uninhibited motion requires spatial

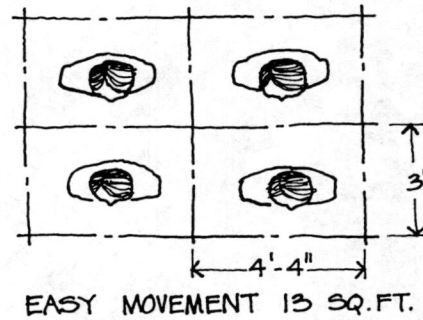

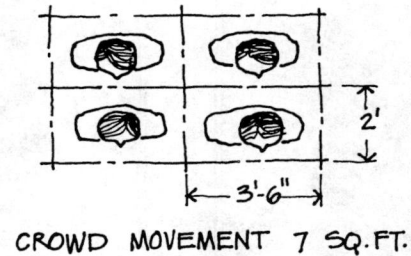

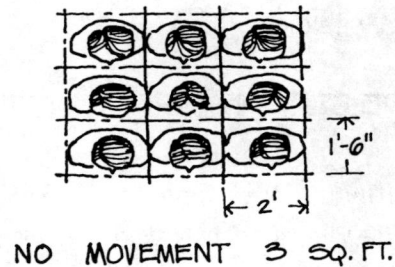

EASY MOVEMENT 13 SQ. FT.

CROWD MOVEMENT 7 SQ. FT.

NO MOVEMENT 3 SQ. FT.

PEDESTRIAN AREA ALLOWANCES

allowances for pacing, sensing, and reacting to other pedestrians. Movement on level surfaces differs from that on stairs or ramps, which require more attention to assure pedestrian safety.

The capacity of walkways varies according to the quality and rate of flow. As a rule of thumb, one-way sidewalks should be no less than five feet wide, and collector walks serving large crowds and two-way traffic, from six to ten feet or more, depending on total capacity and rate of flow. Two-directional flow along a walkway is not as efficient as one-way flow. Pedestrians have a tendency to group themselves into

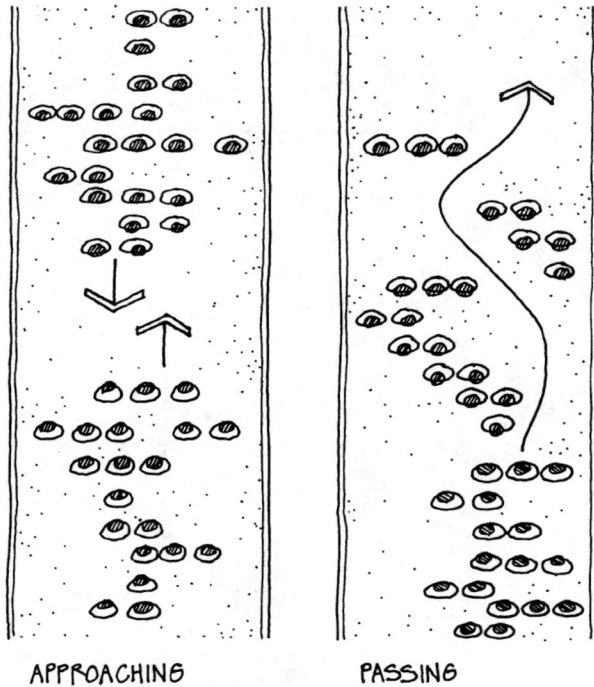

APPROACHING   PASSING

PEDESTRIAN HERD INSTINCT

nonconflicting directional herds, particularly when the rate of flow is nearly equal in both directions.

Crowd characteristics and psychology also play a role in the design of pedestrian ways. For example, 80,000 football fans streaming out of the Rose Bowl after a game will require greater flexibility of movement in dispersing than 3,000 persons leaving Avery Fisher Hall at the conclusion of a concert. Consequently, the designer must allow sufficient space to minimize conflicts which reduce the rate of flow and cause congestion. Each of these aspects must be considered, together with the characteristics and use of the site and its buildings. In a shopping mall, for instance, pedestrians tend to move in linear patterns, criss-crossing diagonally between shops. This requires a greater amount of space to avoid circulation conflicts.

Most people prefer to walk in the line of least resistance. Consequently, foot paths should be designed to provide the most efficient and direct means to circulate between uses on a site. In some situations, however, where activities are primarily ceremonial in nature, the configuration of walks leading to major building entrances, monuments, etc. may be determined more by formal than by functional needs, as at the Capitol Mall in Washington, D.C.

Distance and speed are the major limitations in the design of a site's pedestrian circulation system. Most people are unwilling to walk more than half a mile in performing their routine tasks. Since the average walking speed is about two and one-half miles per hour, pedestrians will walk up to twelve minutes. If the distance is more than half a mile, people prefer to use their cars to reach their destinations.

Many of our older universities, for example, have grown in size beyond the expectations of their founders. Where distances are excessive, students often prefer to drive between facilities, though they may have to walk five or ten minutes to and from their car, leave and re-enter the site, and thereby use more time than the entire trip might require on foot. To avoid this, site planners concentrate high-use facilities within a core comprising a walking radius of one-quarter mile, which circumscribes an area of approximately 125 acres. Low-use facilities, especially those for parking and service, are placed around the perimeter of the core. This reduces walking time within the core, allowing one to traverse the area in about 12 minutes from end to end, and minimizes vehicle-pedestrian conflicts as well.

Walks should be designed to allow people to move safely, independently, and unhindered through the exterior environment. The surfaces

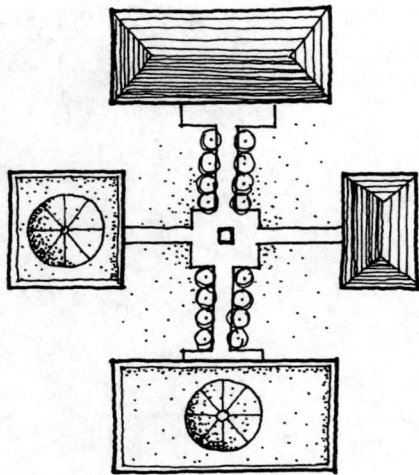

FORMAL

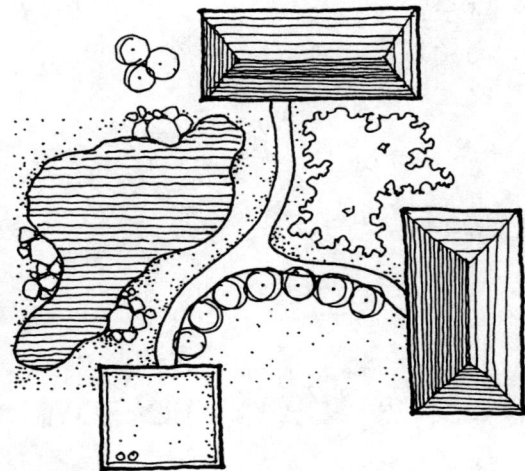

FUNCTIONAL

of walks should be stable and firm, relatively smooth in texture, and have a nonslip quality.

Walking surfaces can be grouped into three types: soft, variable, and hard. Each group has its advantages and disadvantages. The soft surfaces are usually the least expensive initially, but require a high degree of maintenance. They are susceptible to erosion, can withstand only light traffic, and are difficult for handicapped people to walk on. Soft surfaces, such as crushed rock, earth, lawn, river rock, soil cement, and tanbark, are useful for areas where light pedestrian traffic needs a moderately firm surface, such as recreation areas, parks, and nature areas.

Variable surfaces, such as cobblestones, exposed aggregate, flagstone, sand-laid brick, wood decking, and wood disks in sand, have moderate maintenance requirements and moderate to high installation costs. The irregularity of their surfaces and their wide joints make walking difficult for the handicapped. Furthermore, ice and snow can damage the surfaces or be difficult to remove.

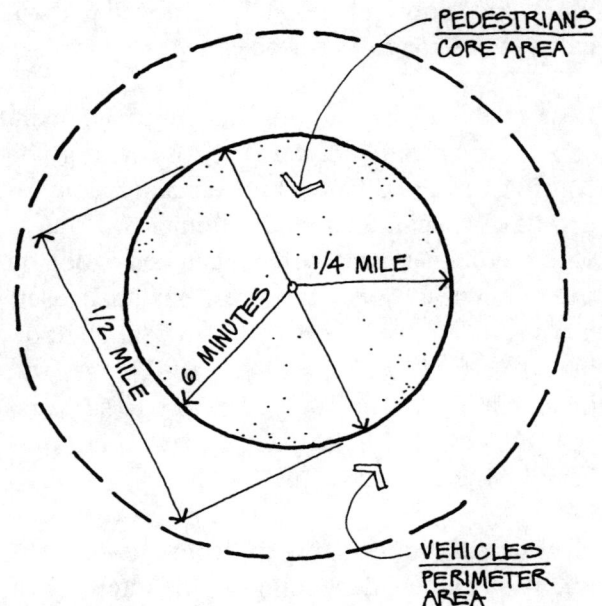

PEDESTRIAN DISTANCE
AND SPEED

Hard surfaces, such as asphalt, concrete, and tile or brick in concrete, are usually the most expensive initially, but their maintenance costs are relatively low. They provide smooth, firm,

WALKS
1:100 (1%) TO
1:20 (5%)

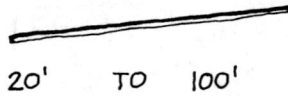

20'   TO   100'

RAMPS
1:20 (5%) TO
1:12 (8%)

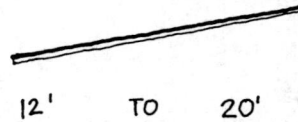

12'   TO   20'

STEPS / STAIRS
1:5 (20%) TO
1:2 (50%)

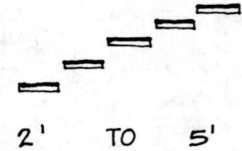

2'   TO   5'

REQUIRED DISTANCES TO RISE OR DROP ONE FOOT VERTICALLY

and regular surfaces for walking or moving wheeled vehicles, including wheelchairs. Expansion and contraction joints, which are less than 1/2" wide and filled, are kept to a minimum, and snow and ice can be removed without extensive damage to surfaces.

Pedestrian paths with gradients under 5 percent (1:20) are considered walks. Those with gradients in excess of 5 percent are considered ramps and have special design requirements. Routes with gradients up to 5 percent can be negotiated independently by the average wheelchair user, but sustained grades of 4 to 5 percent should have five-foot level areas every 100 feet of run or so to allow a handicapped person to stop and rest. However, slopes under 3 percent are preferred, whenever possible.

Walkway lighting may vary from one-half to five footcandles, depending on the intensity of pedestrian use, the hazards present, and the relative need for personal safety.

Changes in grade from street to sidewalk and from sidewalk to building entrances create problems for handicapped persons. Curb ramps not exceeding a gradient of 1:12 (8 percent) facilitate movement over low barriers. Their surfaces

should be nonslip but not corrugated, since frozen water in the grooves can cause ramps to become slippery. Although curbs may be required in certain situations, they should not be higher than 6" to 6-1/2". Curb ramps for the handicapped should be provided wherever pedestrian traffic occurs and where vehicles are parked adjacent to curbs.

Walkways should be pitched laterally about 1/4 inch per foot to eliminate standing water in depressions. For wider walks, the surface should be crowned to slope away from the center.

Stairs may be used where gradients are such that persons in wheelchairs must use an elevator to negotiate the change in level. An exterior stair should have at least three risers in order to avoid pedestrians tripping over steps not easily seen. Stairs with more than four risers should be provided with handrails on at least one side. Exterior stairs typically have 5-1/2" to 6" risers and 15" and 14" treads, respectively. Broad expanses of monumental steps may have 3" to 4" risers with 19" to 17" treads respectively. These should be avoided where pedestrian circulation is heavy, since they are more cumbersome than normal stairs. A useful rule for proportioning conventional steps is that the height of two risers

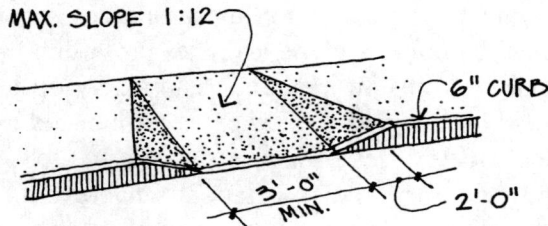

FLARED SIDES

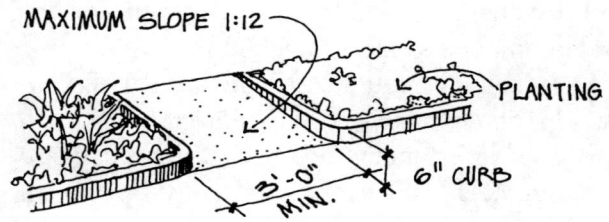

RETURNED CURB

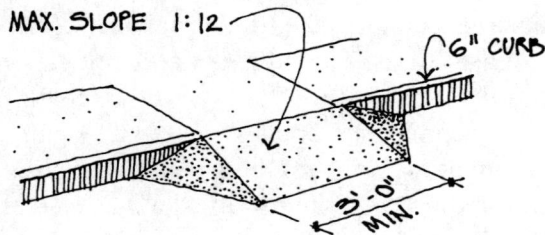

BUILT-UP

PARALLEL

TYPICAL CURB RAMPS

added to the tread dimension should equal 26" to 27". Stairs in heavy public use should never have a gradient over 50 percent, for example, a 6" riser with a 12" tread.

Pedestrian spaces must be sized according to the anticipated intensity of activity. For example, a campus mall in a central location may appear oppressive because of the crowd attending an outdoor concert, while it would seem empty, vast, and lonely in the evening, when most students are no longer on campus. A foot path on the periphery of a commercial center may seem excessively long, while if it were bordered by varied shops and stores and used by many people, it would appear interesting and relatively short. Pedestrian spaces should do more than physically link activities—they must reinforce the composition of the site design, both visually and emotionally.

UTILITIES AND SERVICES

Introduction

The design of a site and the placement of buildings and other developments must consider the availability and location of utilities and services. These include the systems for the supply of water, gas, and electricity, the disposal of waste water and storm water, the control and disposal of flood water, and the connections to various communication networks. Services to collect trash, to protect facilities in case of fire, to distribute mail, and to remove snow and ice also influence site elements. These utilities and services are provided by public agencies or private companies, such as public utility companies which are regulated by the state. Utilities are distributed above or below ground by a variety of devices, including conduits, pipes, channels, tunnels, cables, and wires.

To avoid duplicate circulation channels for vehicles and utilities, and to afford easy access for maintenance, most services are placed within the public right-of-way in streets. These normally include sanitary sewer lines, water mains, gas mains, electric power conduits, telephone conduits, telegraph lines, and cable television lines, and they can be identified by the standard symbols shown below.

Local government agencies regulate the planning and placement of these lines in public streets. On-site distribution, however, is the responsibility of the developer, although the

SYMBOLS FOR SITE DRAWINGS

——————	PROPOSED (NEW) UTILITY
– – – – –	EXISTING UTILITY
—24"SS—	STORM DRAIN (SEWER)
—8"SAN—	SANITARY SEWER (WASTE WATER)
—6"YD—	YARD (SITE) DRAIN LINE
—4" W—	WATER MAIN
—3" G—	GAS MAIN
—E—	ELECTRICAL (POWER) CONDUIT
—T—	TELEPHONE (UNDERGROUND)
—O—	MANHOLE
-(S)-	STREET OR SITE LIGHT
● ○	NEW AND EXISTING FIRE HYDRANT

design and engineering of these services is controlled by codes and the agencies providing the services, to insure the health and safety of its users. The routing of services is influenced by the characteristics of the site, such as its topography, geology, and vegetation. Sewers and storm drains must be sloped, and their layout must respect the natural topography in order to remain below the ground, to drain properly, and to avoid the need for pumping uphill. If, for example, there are sewers in each of two adjacent streets along a site, it is usually preferable to connect to the one which is at an elevation lower than that of the site, even if the length of run to its connection point is greater. Connecting to a line at an elevation higher than the site would require pumping to lift the waste—an expensive and high maintenance solution.

Sanitary Sewers

Sanitary sewers must have sufficient cover to prevent breakage or freezing, but not be so deep that excavation costs become prohibitively expensive. Lines must slope to provide a rate of flow of 2-1/2 to 10 feet per second, in order to transport solid waste. Grades required to accomplish this vary generally between 1/2 and 2 percent. If the topography changes abruptly, sharp differences in elevation can be accomplished by providing drop manholes or pumped lift stations.

The cost of sewers depends on the amount of sewage to be carried, the size and type of conduit, the characteristics of the soil, and the amount of excavation required to provide the necessary cover.

Sizes of sewage lines vary from 8" for lateral connections to many feet in diameter for large trunk lines. Materials used include vitrified clay, cast iron, plastic, and lightweight fiberglass reinforced mortar plastic, in sizes up to four feet in diameter and twenty-foot long sections.

Sewage contains materials which are a health hazard, produce offensive odors, and render water unsuitable. Treatment plants change the composition of waste material prior to its discharge into bodies of water.

Septic Tanks

Where connection to a treatment plant is not feasible, such as in remote rural areas, certain alternatives may be considered. Experience and research have determined that the sewage cesspool is a hazard to health, as well as a nuisance. Where soil is sufficiently pervious and ground water low, it is possible to use a septic tank discharging into an underground drain field.

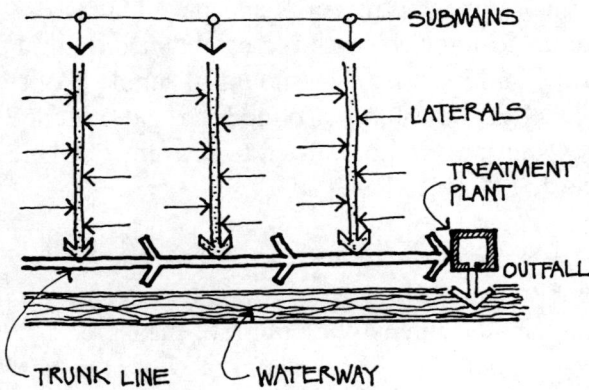

DIAGRAM OF SEWER SYSTEM

Septic tanks change waste into gases and effluent liquid through the action of anaerobic bacteria. The effluent is subsequently rendered harmless by earth leaching, where anaerobic bacteria oxidize all obnoxious components of the waste. The porosity or absorption of the soil is a vital factor in the design of an effluent disposal system.

Septic tanks should be buried and vented, and located as far to leeward of buildings as possible. Their size is determined by the estimated quantity of sewage to be treated. The effluent disposal method is largely dependent on soil conditions, topography, and the amount of waste to be disposed of. Drain fields must be kept 100 feet from any surface water or well, and they should not be heavily shaded or crossed by vehicles.

There are three principal types of systems: leaching cesspools, subsoil disposal beds (underground drain fields), and sand filters. Each system of disposal has its advantages and limitations. Leaching cesspools require a small amount of land regardless of its slope. Their initial cost is low; however, they cannot be located in either semi-impervious or impervious soil. Subsoil disposal beds may be used in any soil except impervious, but not where ground

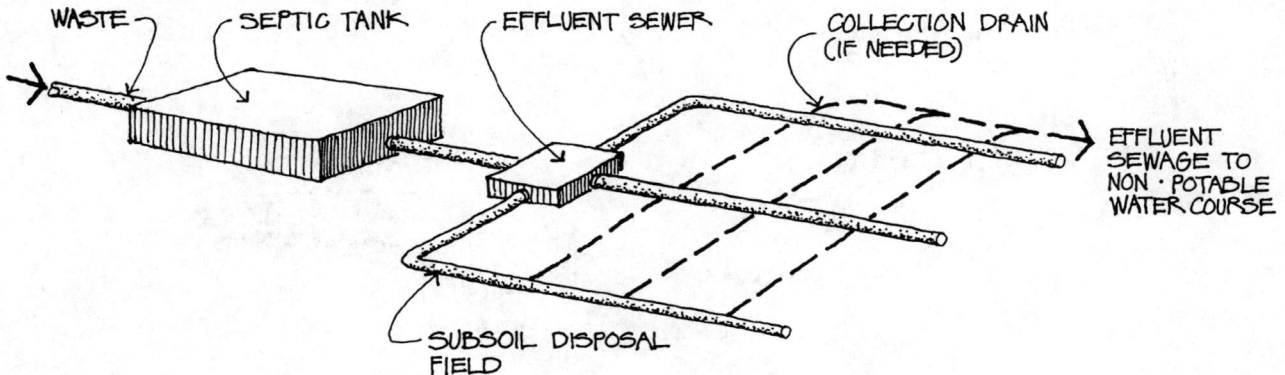

SEPTIC TANK AND SUB·SOIL DISPOSAL FIELD

water is less than two feet below grade. Their cost is greater than that for leaching cesspools, but less than for sand filters. The advantages of sand filters are that they may be used in impervious soils and require a relatively small area. However, they required the use of collection drains, their effluent must be carried to a nonpotable watercourse, and they are relatively expensive.

In developing a new community, the use of septic tanks is more economical than a sewer system with a disposal plant. A disposal plant, however, can be connected to a future public system without any economic loss on the installed sewer pipes. Wherever possible, public sewer systems offer the most economical solution to the disposal of waste water.

Storm Drainage

Rain water always runs downhill, seeking the easiest and quickest flow path. Unless it is intercepted, it can run over site improvements and into buildings, causing considerable damage. To avoid this, water must either be diverted around improvements or collected in drainage devices and conducted underground in pipes to a storm drainage system. Since underground storm drainage systems are expensive, every effort is made to minimize or eliminate their need. This can be achieved by maintaining as much natural ground undeveloped and unpaved as possible, grading carefully to insure gentle slopes and positive water flow, and using swales, ditches, culverts, and other alternatives to piped services wherever possible.

Surface water travels in a sheet across the ground, some of it being absorbed as it runs off. If it moves too rapidly, it will cause the earth to erode. Slopes must be designed in consideration of the volume of water expected, the surface finish of the ground, and the damage that could be caused by flooding. Planted areas should be appropriately sloped in order to drain properly. Water collected by roofs and paved surfaces tends to interfere with the established natural flow and increase the amount of runoff. Consequently, some kind of collection and drainage system is usually required to prevent flooding during storms.

The amount of water that must be carried by underground drainage pipes is based on the anticipated intensity of rainfall, measured in inches per hour. The amount of water that runs off depends on the surface on which it falls: impervious surfaces such as paving will cause

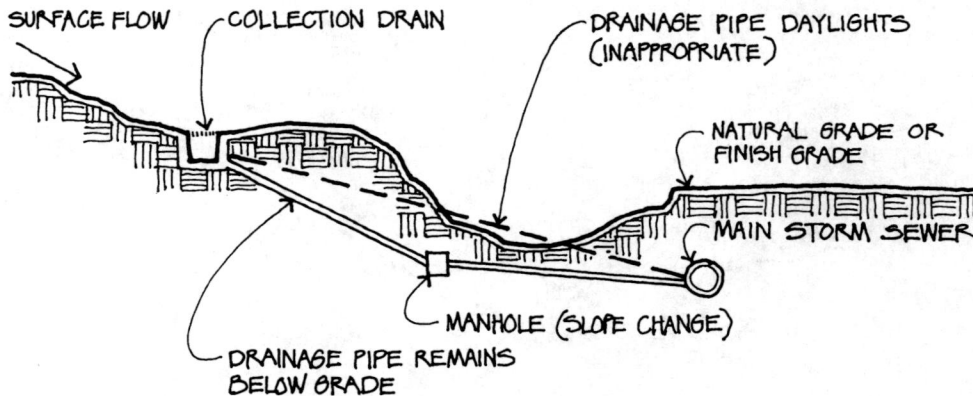

SECTION THROUGH TYPICAL STORM DRAINAGE SYSTEM

nearly all of the water to run off, whereas densely wooded areas may absorb as much as 90 percent of the rain water.

High density developments often require systems with large pipes. These may cause difficulties in developing a grading plan, especially near the lower elevations of pipe runs, where they tend to daylight in order to maintain required slopes to reach their point of connection. Storm sewers must be covered for the same reasons as sanitary sewers. Inlets for these systems are located at various points on a site to allow as much surface flow in open drainage devices as possible prior to being collected for underground piping.

The systems should be designed to minimize both the length of pipe runs and the number of manholes. A storm drainage system is laid out

in plan and elevation from the point of discharge (connection), since this point is most often established by its invert elevation and its location, which may be on or off the site. (See illustration on the following page.)

Factors such as allowable slopes of pipes, location of site drains, topography, and soil conditions may dictate the location of buildings and other improvements on the site. As with circulation systems, the layout of utilities must consider all natural and man-made features as part of the total design of a site. Where subsurface drains are required to remove water from wet ground, to prevent seepage through footings, or to avoid frost heaving on a high water table, perforated pipes are placed in gravel fill and lead into a storm drainage system or natural drainage courses.

Water Supply and Distribution

A water supply system consists of the installations required to obtain, treat, and distribute water to the users. The potential sources for a system include rivers, lakes, and wells.

Transmission mains, including aqueducts, canals, and pipe lines, are used to transport raw water to treatment facilities or directly to

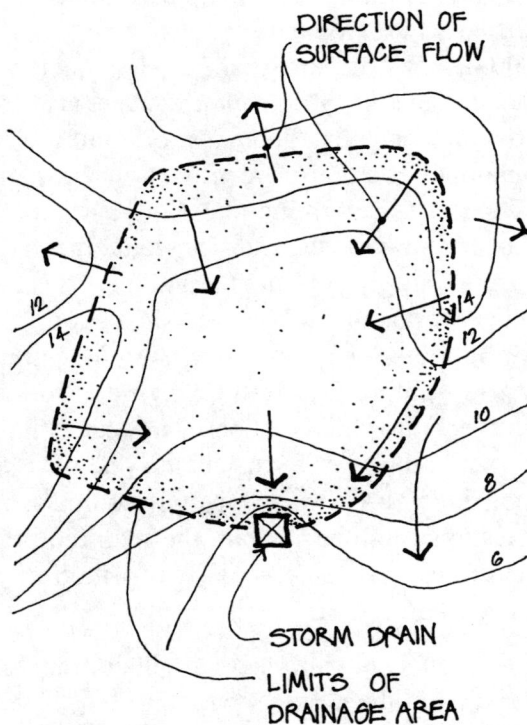

TYPICAL DRAINAGE AREA

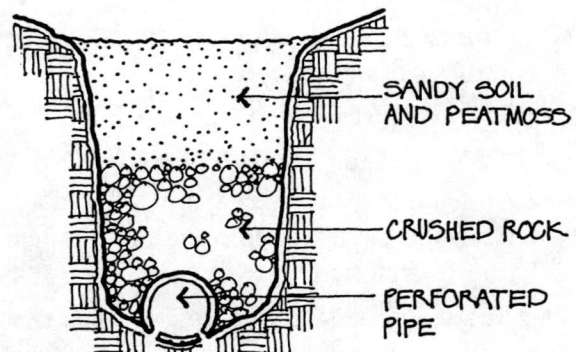

SECTION THROUGH TYPICAL SUBSURFACE DRAIN

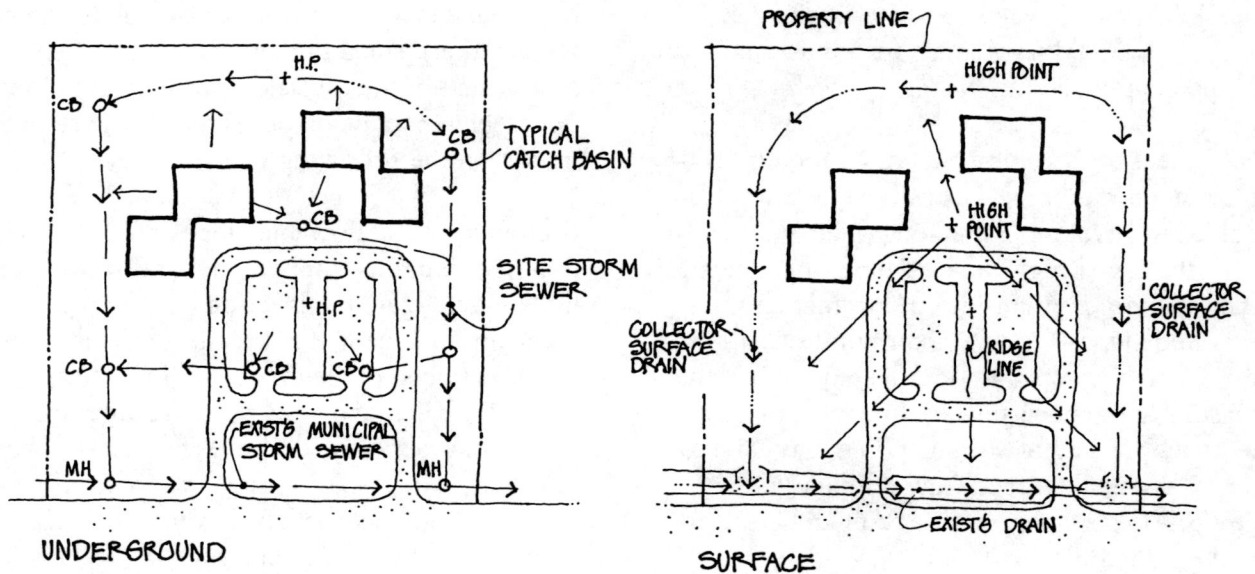

TYPICAL DRAINAGE SYSTEMS

distribution systems if the water is to be left untreated. Most water is treated to meet public health standards for drinking water. Treatment plants may be located either at the water's source, near a river or lake, or at a storage reservoir, from which it is pumped to the points of use. The typical lake or pond water treatment process consists of four general parts:

1. *Settling basin.* Allows large particles in the raw water to settle at the bottom.

2. *Filtration unit.* Water passes through a filter bed of sand and gravel.

3. *Clear water storage.* Water is stored in a clear well, cistern, or storage tank.

4. *Disinfection.* The water is treated with chemicals, including chlorine, to eliminate harmful bacteria.

After treatment, continuing bacteriological examinations are made to ensure proper disinfection of the water at all times. Services to connect the distribution system to the user are provided by either a private or municipal water

company. A public water supply is normally justified where population densities exceed 1,000 persons per square mile. The quantity of water demanded by a development area is based on average and peak demands for domestic and irrigation uses and for fire protection. Estimated demand rates determine plumbing and pipe sizes, pressure losses, and storage facilities necessary to supply sufficient amounts of water during periods of peak usage. Most supply systems distribute water for domestic, industrial, and fire fighting use from a single network. However, many densely populated areas have separate high-pressure fire main systems. Such systems may use the same supply source as the domestic system and obtain the required pressure through pumps. Two basic distribution layouts can be used for water supply:

1. *Central feeder patterns* with lines branching out from the point of entry.

2. *Loop networks* with more than a single point of supply.

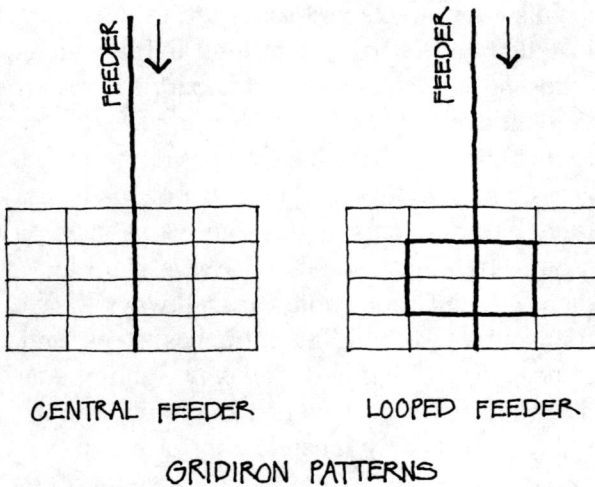

CENTRAL FEEDER   LOOPED FEEDER

GRIDIRON PATTERNS

WATER SUPPLY SYSTEMS

Central feeder patterns are cheaper because runs are shorter, while loop networks are preferred because they avoid pressure drops at the ends of long branches and few, if any, users will be cut off from service when a main breaks.

In order to assure delivery of the maximum instantaneous demand of water at hydrants and buildings, distribution systems are kept under constant pressure. Since they are under pressure, water lines may rise and fall with the slope of the land, as long as positive pressure is maintained throughout the system. Meter and shut-off valves are installed in the line at entry points to buildings or at the property line of an entire development. Fire hydrants are located along roads in order to make all parts of buildings accessible with 500-foot long hoses, preferably from two hydrants. Hydrants should not be closer than 50 feet to buildings in order to maintain safe accessibility in case of fire. They are usually spaced 150 feet apart in urban areas and 600 feet apart in suburban areas.

The minimum diameter for water mains is 6" to 8". Computation of required pipe sizes is complex and depends on demand, pressures, and length of runs. In developments of moderate size, 8" mains are usually adequate. Pipes may be cast iron, wrought iron, steel, plastic, or reinforced concrete, the latter usually for the larger conduits only. Water systems are quite adaptable in terms of locating lines, valves, meters, and hydrants, and less likely to require changes in the design of a site than sanitary or storm sewer systems, which require sloping pipes.

Electric Power Systems

Electric power systems include all of the installations required to deliver electric power from source to consumer. Electric energy is produced by hydroelectric, thermal, or nuclear generating plants. Circuits carry bulk electric power from the source to distribution centers. The distribution system includes substations and transformers to reduce the high voltages of transmission and primary distribution lines to the lower voltage of secondary distribution lines acceptable for use by the consumer. Conventional underground primary distribution systems can be arranged in a variety of patterns including branch, radial, and loop. Loop patterns provide service from more than one direction in case part of the system fails. They are more expensive than branch patterns. Loops are sectionalized by switches or circuit breakers to increase their flexibility and dependability.

In high density areas, conductors may be placed overhead on poles or underground in ducts; in rural and suburban areas they may also be directly buried. Ducts may be made of clay, steel, plastic, or fiber pipe, all of which are generally enclosed in concrete. Underground distribution is considerably more expensive than overhead, but it reduces the potential number of breaks, does not interfere with vegetation, and eliminates the unsightly clutter of power poles. Underground installations are suited to locations where excavation is easy. Where rock or a high water table occurs, costs may become prohibitive, and power poles are virtually mandatory. If one is asked to choose between an overhead and underground installation, the type of soil must be determined before making a recommendation.

Lines are strung on poles except where they enter the building. Poles may be spaced as much as 125 feet apart. Transformers are either exposed and supported on power poles or placed in underground vaults where they must be vented to disperse internal heat.

Site Lighting

The purpose of site lighting is to provide safety and security after dark. Lighting is normally provided in areas used by vehicles and pedestrians, particularly where a dangerous situation would result if they were unlit. Light is especially important at building entrances, at intersections, stairs, ramps, abrupt changes in grade, dead ends, and remote walks. Areas with high crime rates should be well lit to provide some security for those using the facilities after dark. One must place fixtures properly to obtain good lighting. Merely boosting footcandle intensities may be a waste of energy, especially if the fixtures are too far apart or poorly located. The initial cost of fixtures and lamps, the cost of maintenance and replacement, and the quality of lighting provided are important considerations in designing a lighting system.

Visual requirements of drivers are not the same as those of pedestrians, and therefore their lighting should be different. Mercury vapor, metal halide, and high-pressure sodium lamps are more economical and efficient than incandescent and fluorescent lamps, although their color rendition is inferior. Incandescent lamps provide the warmest color, but they are relatively expensive, since much of their energy is dissipated as heat.

Lights over streets and roads are usually installed on 30- to 50-foot high lighting standards, spaced 150 to 250 feet apart. Lamps provide an average illumination of one-half footcandle at local roads and for all parking areas, and one footcandle on major roads and large parking areas. Pedestrian walks may require lighting intensities varying from less than one-half footcandle for walkways to five footcandles for building entrances, steps, and intersections. Mall and walkway lighting standards are between 10 and 15 feet in height. Lamps are usually incandescent or mercury vapor. Low level fixtures, where the light source is below eye level, are sometimes used to illuminate landscaping and pedestrian walks. When these are the primary source of illumination, peripheral lighting to illuminate the immediate surroundings should be provided to produce a feeling of security for pedestrians passing through such areas.

Natural Gas

Distribution systems for natural gas are similar to those for water. Gas is piped underground in branching or loop patterns, its flow is controlled by valves, and the amount consumed is measured by meters. A loop system is preferred, since it provides alternate directions of supply in case of failure. The main problem with the installation of gas lines is the potential danger of leakage and explosion, consequently lines should not be located under or close to buildings except at points of entry, and never in the same trench with power cables.

Materials used for gas pipes are mainly welded steel, although cast iron is still in use in some systems, since this material was originally the standard of the industry. Where natural gas is unavailable, manufactured gas, stored in tanks and distributed in the same way as natural gas, is used. However, this practice is largely limited

LAMP TYPE	WATTAGE RANGE	EFFICIENCY (lumens/watt)	LIFE (hours)	COLORS STRENGTHENED	COLORS DIMINISHED	REMARKS
INCANDESCENT	15-1000	LOW	750-2000	YELLOW, RED, ORANGE	BLUE	GOOD COLOR RENDITION
DELUXE COOL WHITE FLUORESCENT	15-215	MEDIUM	7,500-15000	ALL	NONE	BEST OVERALL COLOR RENDITION
DELUXE WHITE MERCURY VAPOR	90-1000	MEDIUM	10,000-24,000	BLUE, RED, YELLOW	GREEN	GOOD COLOR RENDITION
METAL HALIDE	175-1000	HIGH	7,500-10,500	YELLOW, BLUE GREEN	RED	GOOD COLOR RENDITION
HIGH PRESSURE SODIUM	250-1000	HIGH	10,000-15,000	YELLOW, GREEN, ORANGE	RED, BLUE	POOR COLOR RENDITION

HIGH INTENSITY DISCHARGE

LAMP TYPES AND CHARACTERISTICS

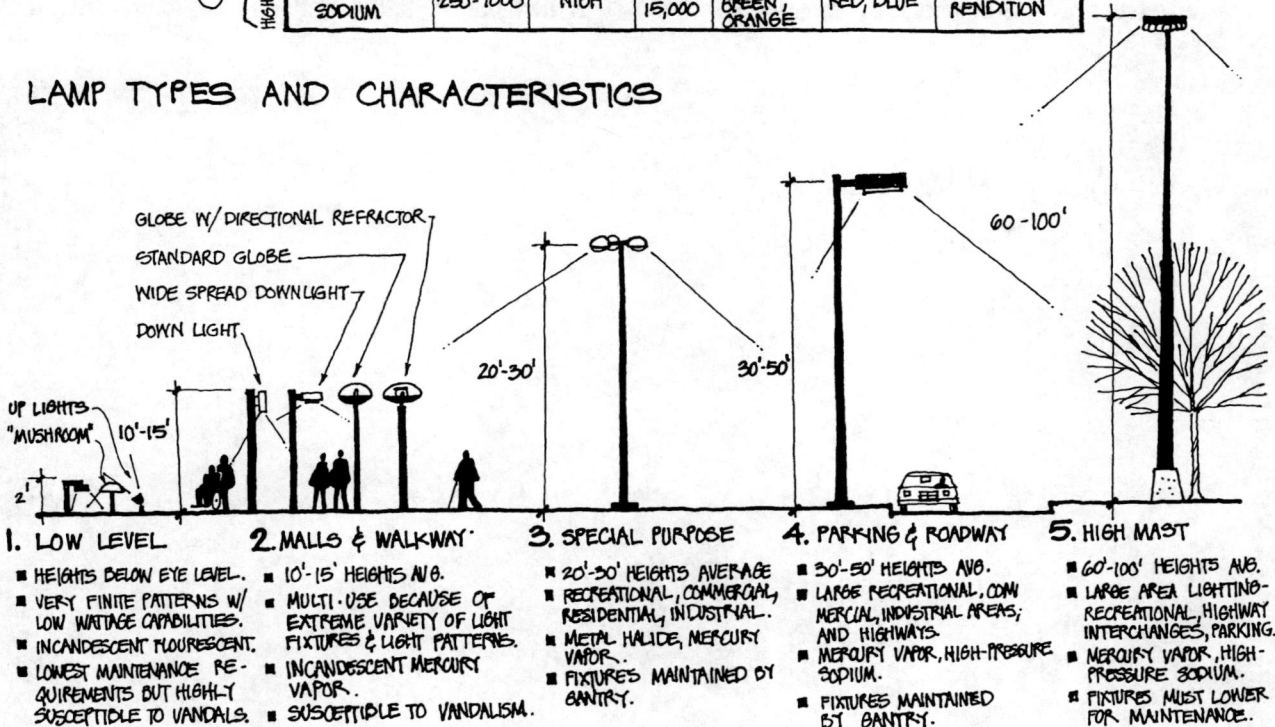

GLOBE W/ DIRECTIONAL REFRACTOR
STANDARD GLOBE
WIDE SPREAD DOWNLIGHT
DOWN LIGHT

UP LIGHTS "MUSHROOM"   10'-15'

20'-30'   30'-50'   60-100'

1. LOW LEVEL
- HEIGHTS BELOW EYE LEVEL.
- VERY FINITE PATTERNS W/ LOW WATTAGE CAPABILITIES.
- INCANDESCENT FLOURESCENT.
- LOWEST MAINTENANCE REQUIREMENTS BUT HIGHLY SUSCEPTIBLE TO VANDALS.

2. MALLS & WALKWAY
- 10'-15' HEIGHTS AVG.
- MULTI-USE BECAUSE OF EXTREME VARIETY OF LIGHT FIXTURES & LIGHT PATTERNS.
- INCANDESCENT MERCURY VAPOR.
- SUSCEPTIBLE TO VANDALISM.

3. SPECIAL PURPOSE
- 20'-30' HEIGHTS AVERAGE
- RECREATIONAL, COMMERCIAL, RESIDENTIAL, INDUSTRIAL.
- METAL HALIDE, MERCURY VAPOR.
- FIXTURES MAINTAINED BY GANTRY.

4. PARKING & ROADWAY
- 30'-50' HEIGHTS AVG.
- LARGE RECREATIONAL, COMMERCIAL, INDUSTRIAL AREAS; AND HIGHWAYS.
- MERCURY VAPOR, HIGH-PRESSURE SODIUM.
- FIXTURES MAINTAINED BY GANTRY.

5. HIGH MAST
- 60'-100' HEIGHTS AVG.
- LARGE AREA LIGHTING- RECREATIONAL, HIGHWAY INTERCHANGES, PARKING.
- MERCURY VAPOR, HIGH-PRESSURE SODIUM.
- FIXTURES MUST LOWER FOR MAINTENANCE.

TYPICAL LIGHTING STANDARDS

to industrial facilities in areas zoned for such purposes. Pipes and mains are sized in relation to the use, pressure, and volume of the gas they transport.

Communication Networks

Communication networks include lines for telephone, telegraph, alarm, and cable television systems. Although trunk lines are generally placed below ground in urban areas, local lines are still strung on electric power poles. In quality planned developments, where the additional cost warrants the placement of these utilities out of sight, they are located underground.

Conclusion

The location of utilities must be considered as a whole, keeping trenching to a minimum, avoiding cross connections, and maintaining horizontal and vertical separations between incompatible systems. Utility layouts must be checked to verify that lines which cross in plan are separated vertically, to avoid conflict.

Where development is relatively intensive and utility systems are numerous and large, placing them in a common tunnel to provide access for maintenance may be cost effective over a period of time. Where buildings are interconnected, utilities, other than natural gas, may be located in common conduits below the structures to simplify access and minimize excavation costs.

The cost to extend utility lines is high. The designer must be able to compare these costs for each particular site condition in order to achieve a workable and economical solution.

LESSON 4 QUIZ

1. Septic tanks change waste into gases and
   A. effluent.
   B. anaerobic bacteria.
   C. potable water.
   D. humus.

2. All of the following statements about storm drainage systems are correct, EXCEPT
   A. storm drains must have sufficient cover to prevent breakage or freezing.
   B. planted areas do not have to be sloped for drainage.
   C. surface drainage is usually more economical than underground drainage.
   D. the amount of runoff depends on the ground surface.

3. A paved parking area should not slope more than
   A. 2 percent.
   B. 1/4 inch per foot.
   C. 5 percent.
   D. 10 percent.

4. Which of the following would probably be the most appropriate and economical surface material for a nature study path?
   A. Crushed rock
   B. Brick pavers
   C. Cobblestones
   D. Concrete

5. Which of the following statements are INCORRECT?
   I. The most desirable locations for vehicular access to urban sites are those close to major intersections.
   II. Public safety is generally the most important consideration in locating points of access to a site.
   III. Grading is minimized when road alignments parallel existing contours.
   IV. Urban sites usually have more choices for locating points of access than rural sites.
   A. I only   C. I, III, and IV
   B. II and IV   D. I and IV

6. Water supply systems
   A. must have a slope of at least 2 percent.
   B. usually have mains at least six to eight inches in diameter.
   C. are invariably provided by a public agency.
   D. usually consist of separate domestic, industrial, and firefighting systems.

7. Sanitary sewer lines should never slope less than
   A. 1/4 inch per foot.
   B. 1/4 percent.
   C. 1/2 percent.
   D. 2 percent.

8. Select the most correct statement.

   A. Non-ambulatory disabilities cause persons to walk with insecurity or difficulty and may necessitate the use of crutches, walkers, or braces.

   B. The farthest distance people are willing to walk, in preference to using their cars to circulate from one area of a campus to another, is approximately one-half mile.

   C. Pedestrians normally circulate along paths laid out in rigid geometric patterns.

   D. School children in an outdoor assembly area would normally require about three square feet each.

9. The best use for a site where an expressway overpass crosses an arterial highway is a(n)

   A. interchange.

   B. shopping center.

   C. roadside rest area.

   D. gas station.

10. Storm drainage

   A. is normally provided by a private utility company.

   B. conduits are usually buried more than 20 feet below finish grade to prevent freezing.

   C. is less critical in urban than rural areas.

   D. conduits are sized on the basis of rainfall intensity measured in inches per hour.

LEGAL AND ECONOMIC FACTORS

LEGAL CONSTRAINTS

Private ownership of property is one of the privileges and rights most treasured by Americans. There are few things that people will defend as fiercely as their land. But property rights are not absolute. As our society has grown increasingly populous and complex, most communities have found it necessary to place limitations on the use of property, in order to safeguard the public's health, safety, and welfare. Thus, every owner of land must comply with zoning ordinances, building codes, and a myriad of other regulations.

Originally, land meant not only the surface, but all the ground beneath it and the air above it. Therefore, an owner could develop and use buildings or other structures on and below the surface for his or her private good.

Present property rights, however, often limit the use of the land below and the air above the surface, depending on mining and water laws and private and public air rights above certain heights. Surface rights are generally limited by government regulations or private restrictions or both, to prevent an owner from developing his or her land in a way that might deprive neighboring owners of the reasonable use and enjoyment of

UNLIMITED TITLE TO AIR SPACE
ABOVE SURFACE OF LAND

SURFACE RIGHTS / LIMITED BY
LAW OF NUISANCE
SUBSURFACE AND CONTRACT
RIGHTS

LIMITED BY
RIGHT OF ADJOINING
PROPERTY TO LATERAL
CENTER SUPPORT & BY CONTRACT

PROPERTY RIGHTS AFTER REVOLUTION AND BREAK FROM ENGLAND

AIR RIGHTS SHARED W/ PUBLIC

AIR RIGHTS OWNED BUT RE-
STRICTED BY ZONING
HEIGHT REGULATIONS

SURFACE RIGHTS / LIMITED BY
GOV'T REGULA-
SUBSURFACE TIONS & PRIVATE
RIGHTS REGULATIONS

LIMITED BY
MINING & WATER
LAWS OF GOV'T
CENTER AND PRIVATE
RESTRICTIONS

PRESENT PROPERTY RIGHTS

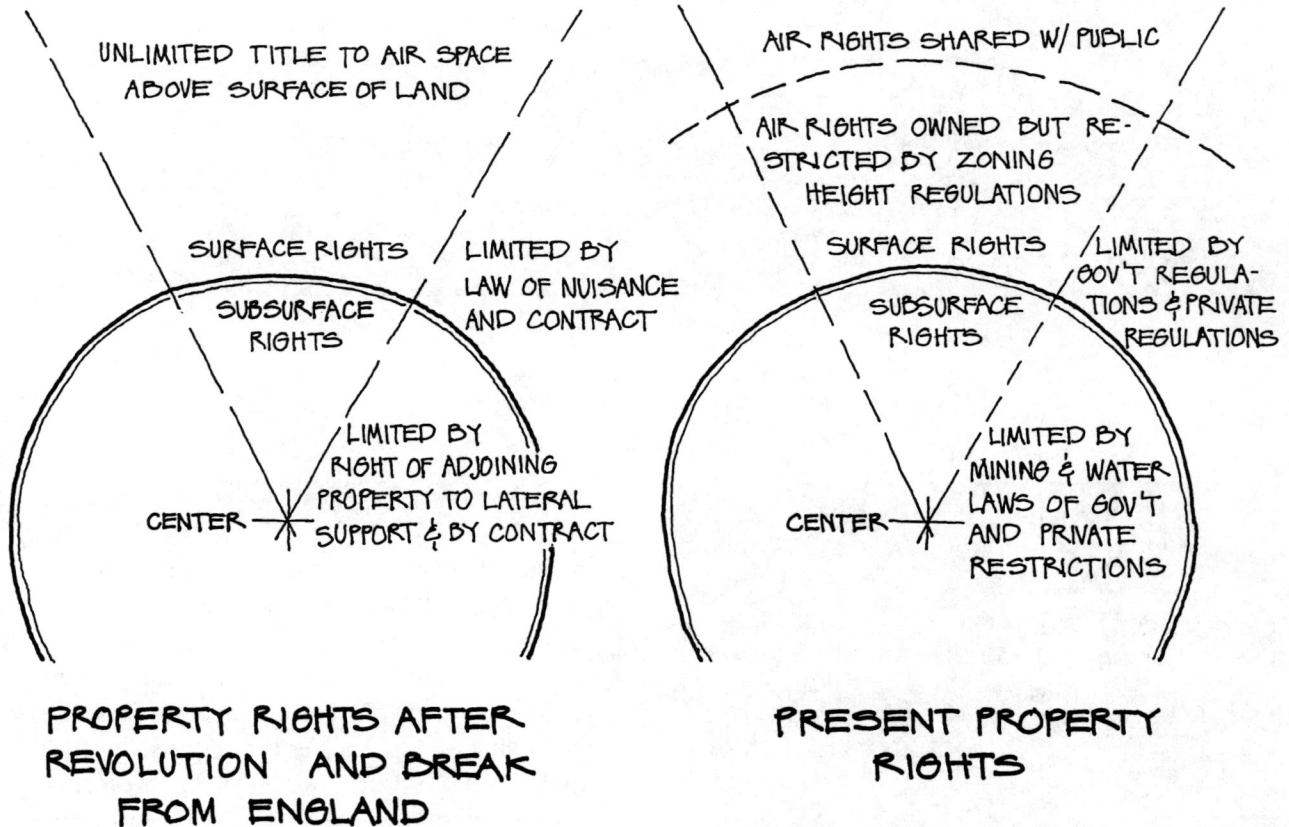

their property. For example, an owner may be prevented from creating hazards, such as fire, noise, air and water pollution, and danger to aircraft. Consequently, a height restriction may be imposed on property located in the landing pattern of aircraft, to avoid interference with their takeoff or landing. Or, the owner may be prevented from developing a property for uses which are morally or psychologically objectionable. Generally, however, tangible activities, such as changing the contours of the land or constructing a building, are more likely to be regulated than are the more intangible matters, such as peace of mind and aesthetics. Public controls are normally exercised through the enactment and administration of zoning ordinances. Additional controls have sometimes been imposed to preserve historic buildings and sites, as well as improve the quality of the environment.

ZONING

General

The first American zoning ordinance was enacted in New York City in 1916 to limit the size and shape of new skyscrapers so that the adjacent streets would not become permanently shaded canyons. Most zoning statutes in the 1920s dealt with physical development. They divided cities into districts of different uses, with uniform regulations for each. For example, residential districts permitted only residences, commercial districts only commercial activities, and so on. Major categories were further divided; for example, industry was divided into heavy manufacturing and light manufacturing. Residential use was divided, into single-family, two-family, and

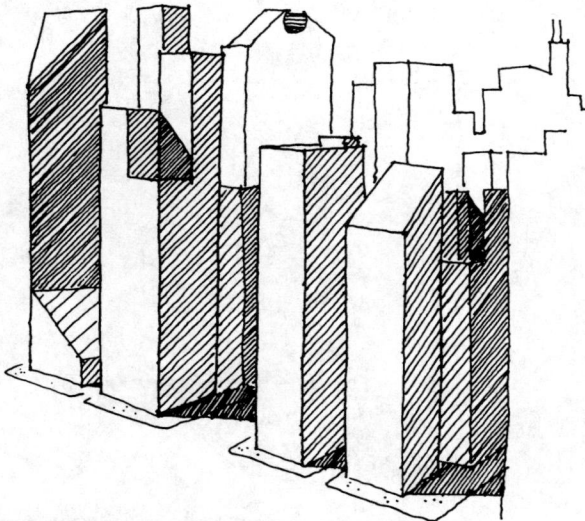

SHADED URBAN CANYONS

multiple-family dwellings. Homogeneous districts were based on the idea that differing uses within a district would lower property values. Multi-use districts, sometimes called *cumulative zoning*, allowed residences in commercial zones and residential and commercial uses in industrial zones.

The ordinances of the 1920s also regulated the height and bulk of buildings and setback lines. The intent of these acts was to allow the owner to develop the land as he or she wished, as long as the specific restrictions of the ordinances were not violated. These laws authorized, but did not compel, local authorities to control development decisions. They did not offer incentives to owners to undertake desirable development, but rather, attempted to avoid undesirable development.

Since that time, the ordinances have changed considerably; instead of prohibiting poor planning, they now encourage, and sometimes even compel, desirable planning.

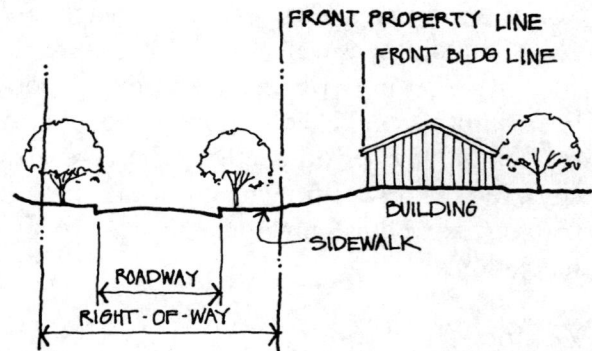

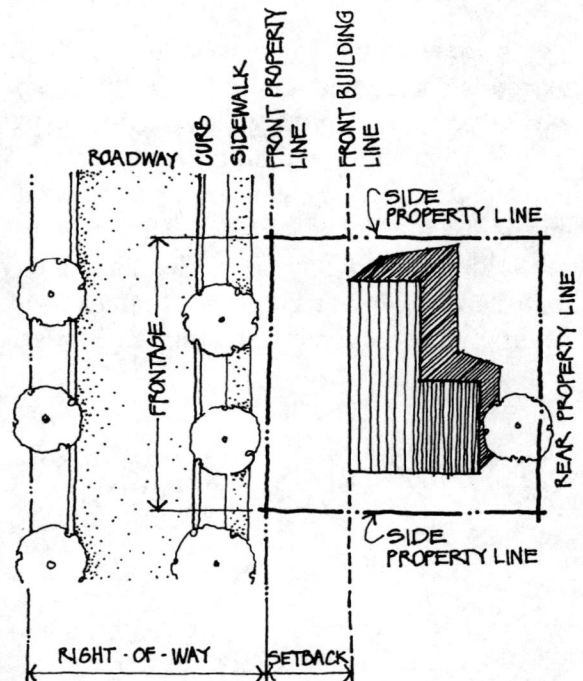

TYPICAL DEVELOPMENT STANDARDS

The enactment of the Model Land Development Code recognized aesthetics, environmental problems, and the preservaion of historical sites as planning and development factors. In New York City, developers commonly add plazas at the ground level of office towers, in return for permission to erect taller buildings. For example, IBM was allowed to build five additional floors, in return for creating a tree-filled atrium at the foot of its Madison Avenue building.

Zoning ordinances often restrict the height and size of buildings, as well as their location on the site. They may prescribe setbacks from property lines, limit percentage coverage of the lot area, restrict the number of dwellings per acre (density), require a specific amount of off-street parking, as well as numerous other possible regulations.

Zoning Envelope

The volume within which a building may be placed is sometimes referred to as the *zoning envelope*. This is an imaginary, tent-like space inside of which the building may be placed in any location, so long as it does not penetrate any of the imaginary surfaces. The drawing below depicts a highly simplified zoning envelope. Most of these have more complex shapes determined by several interacting standards and restrictions.

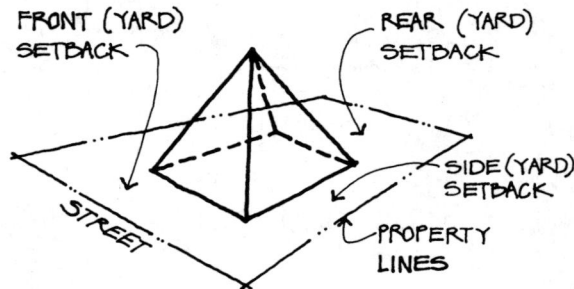

THE ZONING ENVELOPE

Setbacks and Yards

A zoning ordinance often regulates the distance between a street and a building, as well as between buildings. The main purposes of such restrictions are to provide building interiors with natural light and ventilation, inhibit the spread of fire from one structure to the next, and minimize conflicts between street traffic and off-street activities. These ordinances also allow for

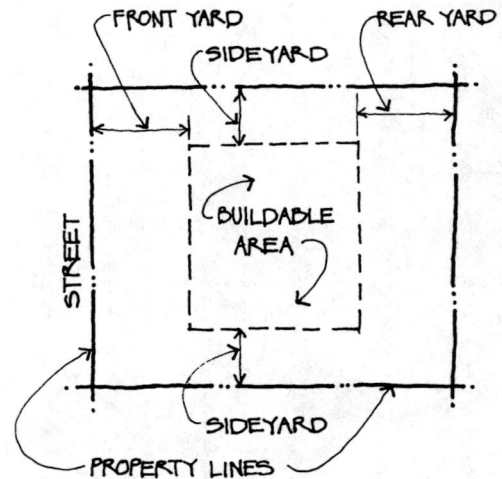

**TYPICAL YARDS
SINGLE FAMILY DISTRICT**

future street widening and create space open from the ground to the sky. The regulations for yards and other setbacks establish the base of the zoning envelope (the ground area within which construction may occur). The drawing below shows a lot for a single-family house and the various yards referred to in zoning ordinances.

A setback is the horizontal space adjacent to a property line into which a structure may not project. Setbacks provide a sense of openness, as well as light and air. They may also be required for off-street parking, or they may be primarily for aesthetic reasons. In some instances, setbacks are established as a function of the building's height using a formula which requires taller buildings to be further back from property lines, in order to assure a minimum amount of openness to the sky. Most ordinances allow paved parking areas within setbacks; however, they may require trees at specific intervals or low walls at property lines in order to screen the rows of parked automobiles.

$$S = H \div \tan\alpha$$

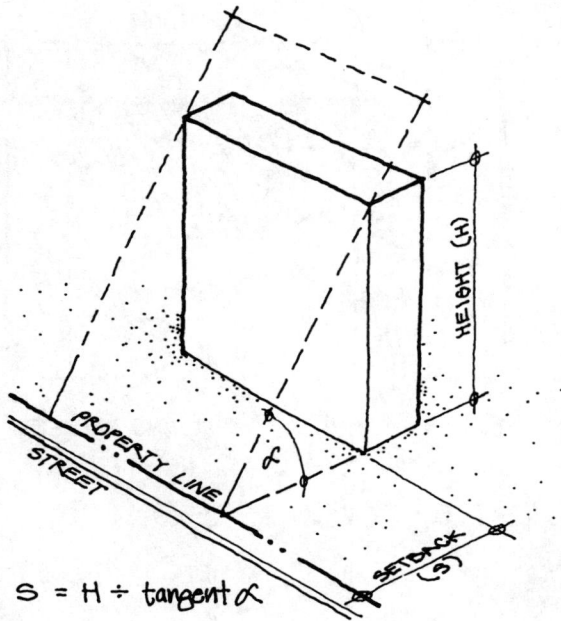

SETBACK AS A FUNCTION OF HEIGHT

Height Limitations and Variable Setbacks

Zoning ordinances may limit the number of stories in a building, its height in feet above street level, or both. The height is usually measured from grade, which may be defined in various ways, depending on the local zoning ordinance. For example, grade may be defined as the lowest adjacent ground elevation, the average adjacent ground elevation, or perhaps some other elevation. Thus, the way in which grade is defined may affect the number of stories permitted, particularly on a sloping site.

Some zoning ordinances contain variable height and setback requirements, either in place of absolute limits, or in addition to them. For example, in the following diagram, no portion of a building may be placed closer to the street than an imaginary plane inclined at an angle of

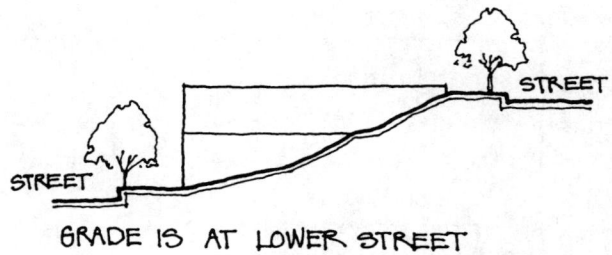

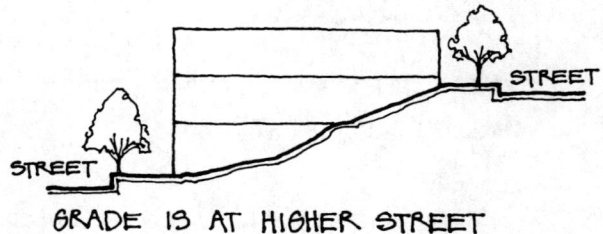

GRADE IS AT LOWER STREET

GRADE IS AT HIGHER STREET

ALTERNATE INTERPETATION OF GRADE

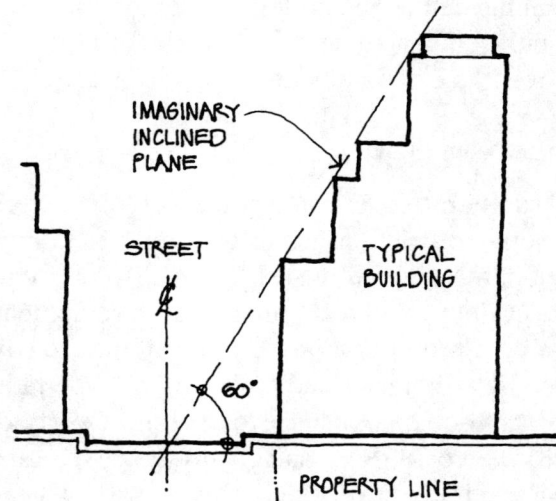

SETBACK BASED ON INCLINED PLANE

60 degrees with the street and extending upward from the center line of the street. A control of this sort encourages a "ziggurat" building profile.

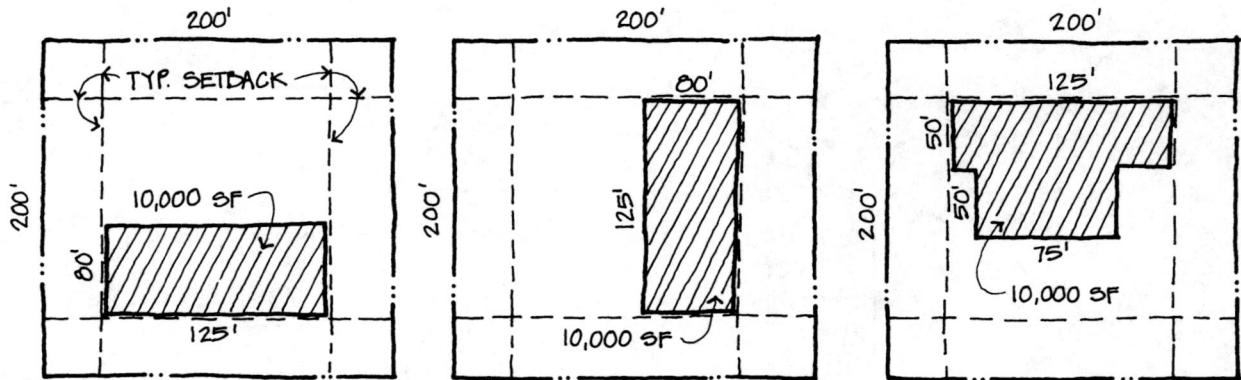

IF MAXIMUM LAND COVERAGE = 25%   .25 x 200' x 200' = 10,000 SF BUILDING FLOOR

THREE OPTIONS FOR LAND COVERAGE

Height limitations are more common in residential zones than in commercial or industrial zones. Crowded urban areas, such as New York City, limit the number of stories in order to control the population density and the resulting traffic, and to retain a certain amount of sky exposure.

Land Coverage

In addition to the basic zoning envelope, there are other restrictions which determine how large a building may be placed on a site. These controls, which involve the amount of development, limit the proportion of the site that may be covered by buildings (land coverage) and the ratio of total usable floor space to total site area (floor area ratio or FAR). Land coverage is expressed as a maximum percentage of total available land area that may be covered by a building or buildings. The open land may be used for surface developments, such as parking, plazas, recreation and other types of landscaped spaces, and level or depressed courtyards. The purpose of these restrictions is to encourage the retention and development of open spaces, to enhance the environment through the admittance of light and air, and to provide planted areas to relieve the hard surfaces of buildings, sidewalks, and streets.

Floor Area Ratios

The floor area ratio (FAR) is the ratio of the floor area of a building to the total area of the site. The purpose of an FAR ordinance is to control the amount of site development and to restrict the bulk of a building, in order to encourage openness, light, and air, especially in urban areas.

Thus, a floor area ratio of 2.0 would permit 40,000 square feet of floor space on a site of 20,000 square feet (2 × 20,000). A ratio of 2.5 would permit 50,000 square feet of floor space on the same site. The drawings on the following page show three options that would be available to the owner of a lot having a maximum FAR of 4. In each of the three illustrations, the lot size is the same: 100,000 square feet. Thus in each case, the maximum floor area allowed is 400,000 square feet. In figure A, the owner has covered the entire site with a four-story structure containing 100,000 square feet per floor. In figure B, half of the site has been covered.

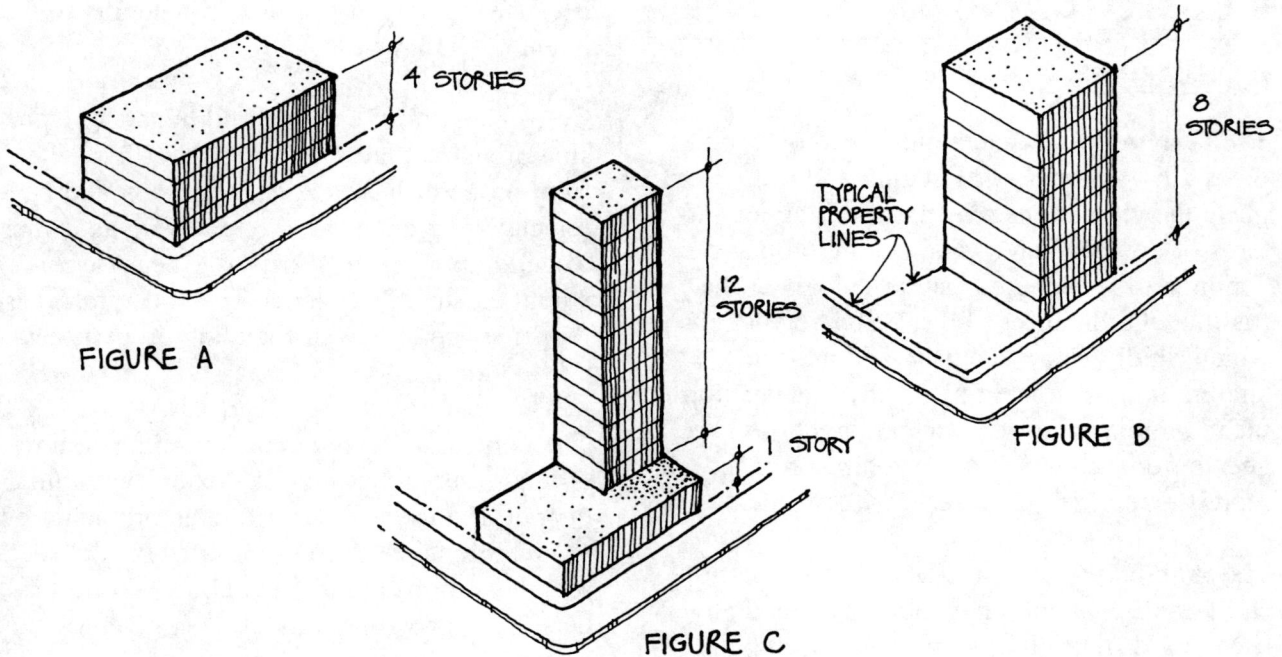

FIGURE A

FIGURE B

FIGURE C

THREE OPTIONS WHERE FLOOR·AREA·RATIO IS 4.0/LOT IS 100,000 S.F.

Since each floor has 50,000 square feet, the building can be eight stories. In figure C, the entire site is covered with a one-story structure, using up 100,000 of the 400,000 square feet of floor area allowed, and the remaining 300,000 square feet have been put into a twelve-story tower, with each level containing 25,000 square feet of usable floor area.

In all of these examples, floor area may be either the net usable space, excluding stairs, elevators, and other similar spaces, or the total gross area of the building, depending on the ordinance.

Off-Street Requirements

Many cities require an owner to provide a minimum number of off-street parking spaces for a building's tenants and visitors. For residential buildings these requirements are usually expressed in terms of parking spaces per dwelling unit. If dwellings are small, one space may be required; for larger units, however, two spaces are normally a minimum. For office buildings in commercial zones, the requirement is usually stated as one space for a specified amount of usable floor area. For example, if the requirement is one parking space for every 500 square feet of usable floor area and the building contains 50,000 square feet, the owner must provide parking facilities for no less than 100 cars (50,000 ÷ 500) within the limits of his or her property. In some cases where space is limited, it may be possible to satisfy the parking requirement on a separate site, provided it is located within a prescribed distance from the building site. Some districts also require a certain amount of site area for loading and unloading of service vehicles, particularly in the case of hospitals, hotels, and institutions.

FLEXIBLE ZONING

General

The purpose of flexible zoning is to overcome the rigidity of traditional zoning and to make the regulations relevant to changing patterns of development. Most zoning ordinances continue to reflect the basic principles of the traditional ordinances of the 1920s: regulations which rigidly define the way land may be used and limitations concerning its physical development. More recently, certain modifications have been introduced which make zoning ordinances more flexible, while preserving their intent.

The *conditional use*, for example, is a departure from traditional zoning, which prohibited any uses in a district other than those specifically allowed by the ordinance. Other significant deviations from traditional zoning concepts include the *planned unit development* (cluster concept), the *floating zone, incentive (bonus) zoning*, and *contract zoning*. Together, these devices are sometimes called *flexible zoning*.

Variances and Conditional Uses

Because even the best zoning ordinances may cause an unintentional hardship to owners of specific land parcels, most cities have established boards which have the authority to grant exceptions to or deviations from the precise terms of these ordinances. These exceptions are called *variances*. Theoretically, a variance is granted when the literal application of an ordinance would cause an undue hardship in the proposed development of a site. For example, a site may have a width of 280 feet in an area where the zoning ordinance specifies the minimum to be 300 feet; however, the site exceeds the minimum lot size by 20 percent. Under these circumstances, a zoning board might grant a variance reducing the lot width requirement

because the property conforms generally to the intent of the law.

Conversely, if all existing buildings along one side of a street are set back 20 feet from the front property line, a zoning board would be reluctant to grant a variance allowing an owner to build up to the property line, because this would create a detrimental visual contrast, easily perceived by the surrounding property owners and the public.

The purpose of a *conditional use* in a zoning ordinance is to provide for flexibility within a district. If a use is described in an ordinance as a "conditional use," it is permitted only if specified conditions are met, a public hearing has been held, and approval has been given by the local governing body. A conditional use is normally granted if it is considered to be in the public interest. A school serving local residents, for example, may be permitted in a residential zone, providing it conforms to certain criteria for traffic, pedestrian walks and crosswalks, and noise control. Under such circumstances a zoning board may grant the conditional use of a site subject to restrictions for the protection of adjacent property owners. The granting of a conditional use or special-use permit does not, however, change the zoning of the particular parcel of land. If the development is abandoned, the conditional use would no longer apply and the property would revert to its original district designation.

Rezoning

The only alternative available to a landowner who cannot meet the requirements for a conditional use permit is to seek rezoning of his or her property. Rezoning, however, can cause hardships to neighboring property owners. Under the strict interpretation of a zoning ordinance, a choice may have to be made between

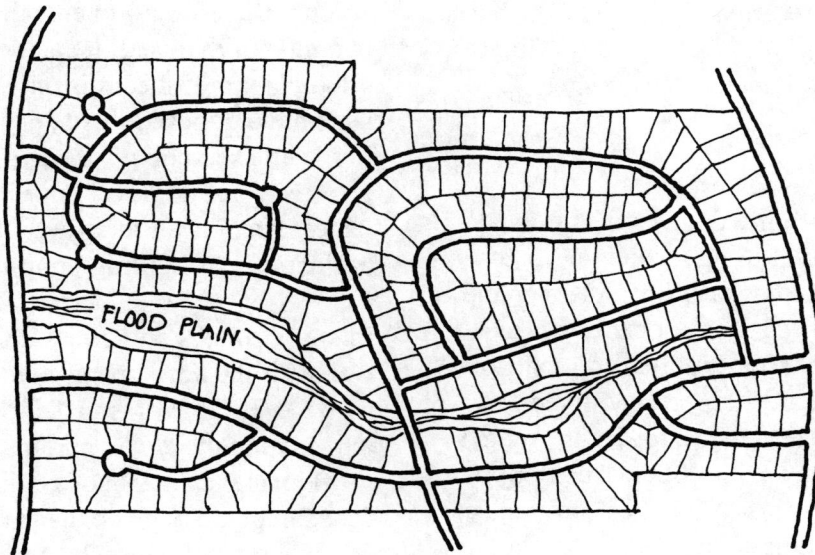

CONVENTIONAL DEVELOPMENT / TRADITIONAL ZONING

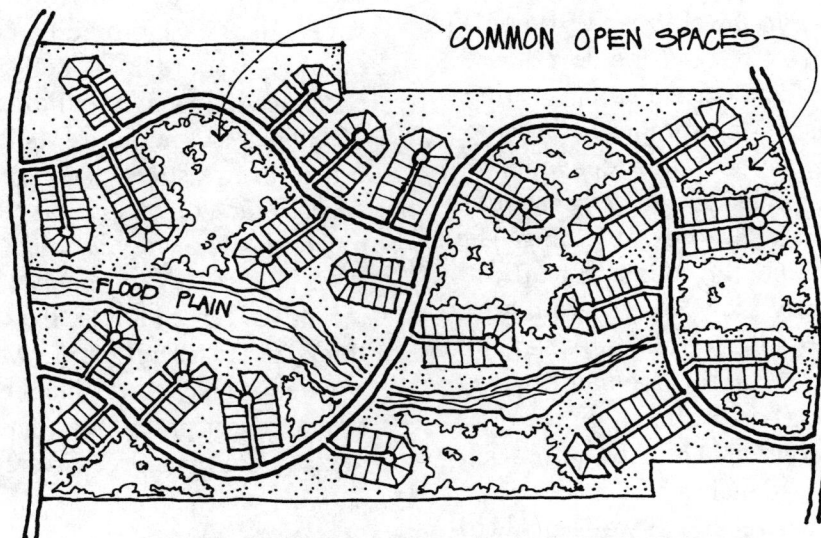

CLUSTER DEVELOPMENT / FLEXIBLE ZONING

FLEXIBLE ZONING

the interests of the landowner and those of the neighboring property owners. Rezoning small individual lots results in *spot zoning*, which may alleviate an owner's hardships. There are times, however, when rezoning is accomplished through political manipulation, rather than for legitimate reasons. For example, rezoning a parcel of property from residential to commercial might appear to have a reasonable basis, while the real reason may be to increase the value of the property for the benefit of its owner.

Contract Zoning

An agreement between a developer and local government to restrict usage or height, or to provide additional setbacks or buffers, over and above what is required by the ordinance, in return for certain benefits, is called contract zoning. For example, a developer may agree to additional restrictions in return for being granted approval of a conditional use. Such restrictions exceed the requirements of the local ordinance and are legally binding.

Contract zoning gives the local governing body power to rezone land or issue special permits granting permission to develop land for non-conforming uses in exchange for a developer's commitment to perform certain compensating acts. In San Francisco, for example, a developer must pay for the construction or renovation of a certain number of housing units whenever he or she plans to erect a new office building. The purpose is to create new dwellings for office workers in an already crowded area with housing shortages. Should the developer balk, he or she must contribute a certain amount of money to a housing bond program, based on the area of new office construction. Other promised acts or conditions might include noise abatement, traffic control, or the erection of walls and landscaped buffer zones. As an alternative, the developer may be allowed to contribute the necessary funds to the local authority in lieu of performing the work himself.

Bonus or Incentive Zoning

Traditional zoning is inflexible in the sense that owners are forbidden to develop their property contrary to the zoning ordinance. Traditional

zoning, therefore, prevents the worst from happening, but that is all it can do. It cannot assure good planning since it is only a restraint. Architects have opposed the negative aspects of zoning, while searching for ways to make it a more positive force in development. Gradually, ordinances have been modified in order to reward builders for benefiting communities.

In some cities, zoning requirements may be waived if the developer provides bonus features, as in the IBM building in New York. This is often attractive to the developer, because not only can the floor area or height of a building be increased, for example, but the bonus features may provide amenities which make the project more desirable for tenants, thereby increasing rents.

Incentives can be given for a variety of reasons: street widenings, providing unobstructed views (as along a shore line), inclusion of theaters and retail space in office buildings, provision of walkways for public use (such as a pedestrian bridge over a street), and preservation of open space. Since open space is the most common objective of incentive zoning, allowing a greater floor area ratio is probably the most prevalent incentive. For example, the developers of the Bankers Trust Building in Manhattan were allowed greater tower height and floor area in exchange for providing a large elevated open plaza and a two-level covered arcade of shops.

OTHER REQUIREMENTS AND RESTRICTIONS

Open Space

To architects, open space implies an aesthetically pleasing area or one that is usable for recreation. Various techniques have evolved to preserve or provide open space. For example, greenbelts were developed in England to create

INCENTIVE ZONING :
A STREET - LEVEL ART GALLERY
IN EXCHANGE FOR A LARGER F.A.R.

buffer zones between developed areas. *Cluster zoning* allows more concentrated density, usually in the form of high rise multiple dwellings, in exchange for common open space. The resultant overall density, however, may not exceed the allowable limits; cluster zoning simply provides for recreational or other open areas without the need for additional land.

Localities sometimes create special districts for the express purpose of preserving open space. For example, Palo Alto, California created such a district in 1972, which limited its undeveloped foothills to open space uses for public recreation, enjoyment of scenic beauty,

and conservation of natural resources. Single-family houses were allowed to be built, but only on very large lots with minimum land coverage.

Aesthetics and Control

It has always been understood that police power could be used to restrict development rights for the protection of public health, safety, and welfare. But could property rights be limited in order to create a more aesthetic environment? Yes, such limitations are legal; some communities have enacted regulations to promote design unity in residential developments, where compliance with aesthetic standards is determined by an architectural board. These bodies have the power to review the design of structures, and they may legally withhold approval if a building design fails to conform to certain minimum standards of appearance. The purpose of such controls is to avoid unconventional structures that would affect the value of surrounding properties or the general welfare of the community. In most instances where the authority of review boards has been challenged, their decisions have been upheld by the courts.

Environmental Constraints

The National Environmental Policy Act of 1969 requires the preparation of an *Environmental Impact Statement* (EIS) for every project which involves "major federal action" having an impact on the quality of the human environment. Any physical development affecting an area's topography, vegetation, geology, wildlife, climate, air quality, etc., must be assessed before being approved, in order to determine its impact on the existing environment.

Similar requirements have been enacted into law by many states and municipalities in order to protect their environment. These laws require that a detailed statement be prepared outlining the environmental impact of the proposed project; whether its implementation would have any adverse effects on the environment; whether there are alternatives to the proposed development; and whether the development involves any irreversible commitments of resources in its implementation, such as the permanent rerouting of a natural water course.

The purpose of an EIS is to provide a basis for regulatory agencies and other interested parties to review the implications of a project and to suggest methods to minimize adverse effects on the site and its surroundings. The statements in themselves do not determine whether or not a project is approved, but they are designed to aid the governing body's final decision.

Additional environmental legislation has been enacted by the federal government and some states to preserve places of natural beauty such as coastlines, lakeshores, and scenic canyons. Development may also adversely affect urban areas; for example, a highway within a city can displace people and harm the quality of urban life. Major projects, such as large office buildings, schools and colleges, hospitals, and housing complexes can also damage the urban environment, and therefore an environmental impact assessment for such projects may also be required.

Opponents of an ordinance and those who wish to contest a permit decision often go to court. In land-use litigation, a person whose property is directly affected can challenge the ordinance or official decision. While some of those suits have been successful, most judicial challenges to local decisions have failed. In upholding a challenged ordinance, the Supreme Court summarized the general feelings of the American public in regard to improving the quality of life with this lofty statement: "A quiet place where yards are wide, people few, and motor vehicles restricted are legitimate guidelines in a land-use project addressed to family needs. The police power is not confined to the elimination of filth, stench, and unhealthy places. It is ample to lay out zones where family values, youth values, and the blessings of quiet seclusion and clean air make the area a sanctuary for people." That statement by the nation's highest court will, no doubt, make constitutional attacks on local development control more difficult in the future.

Easements and Rights-of-Way

An *easement* is the legal right of a government or a landowner to make use of the property of another landowner for a particular purpose, for example, the right to traverse a neighbor's land to gain access to one's own, which is known as a *right-of-way*. Easements exist for a variety of purposes; one common type of easement is the right to use privately-owned land for sewers, utility lines, and roads. Another type of easement permits the construction of an access road or drive between two properties for the use of both owners, who usually sign a written agreement to that effect. However, a written agreement is not always necessary; such easements have been legally upheld if there is a verbal agreement and continuous use over a long period of time.

Easements and rights-of-way are also created by condemnation through the power of *eminent domain*, a court proceeding usually initiated by a public body desiring to acquire a right or title to private property. Public utilities may gain access to properties in that way, to construct power lines for the benefit of the public.

Public rights-of-way may be created by the long and continuous use of a pathway over private land, where the owner fails to take action to prevent such use of his or her land.

After a certain period of time, the owner may lose the right to contest the usage, and a public right-of-way may be created without any additional action. More often, however, public rights-of-way are created by condemnation. For example, when land is taken by condemnation for the construction of a road or power line, the taker acquires an easement rather than complete ownership. Such an easement can, in the case of condemnation for a highway, create a public right-of-way.

Drainage and Surface Waters

Construction frequently affects natural drainage courses, which can cause problems involving the collection and disposal of surface water. Consequently, the designer must be aware of the legal ramifications of surface water drainage. There are three basic rules in this regard.

1. The "common enemy" rule assumes that each landowner may treat surface water as a common enemy, with the right to deal with it as he or she pleases, disregarding the effects on neighboring property.

2. The "civil law" rule holds a landowner who interferes with the natural flow of water liable for any damage to his neighbor as a result of his actions in altering existing drainage patterns. Both of these rules are absolute: the first gives an owner absolute freedom to deal with surface water; the other creates absolute liability for any diversions or changes in the natural flow.

3. The "reasonable use" rule has become more acceptable and widely applied because it recognizes neither absolute liability nor total freedom to deal with surface water. Where the "reasonable use" rule is in effect, each case is decided on its own facts. The landowner affecting the water flow must act reasonably, the determination being based on these considerations:

   1. Is there a reasonable necessity for such drainage?

   2. Has care been taken to avoid injury to the land receiving the water?

   3. Does the benefit to the drained land outweigh the resulting harm?

   4. Is the diversion of flow accomplished by improving the natural system of drainage or has a feasible artificial drainage system been installed?

PROPERTY A

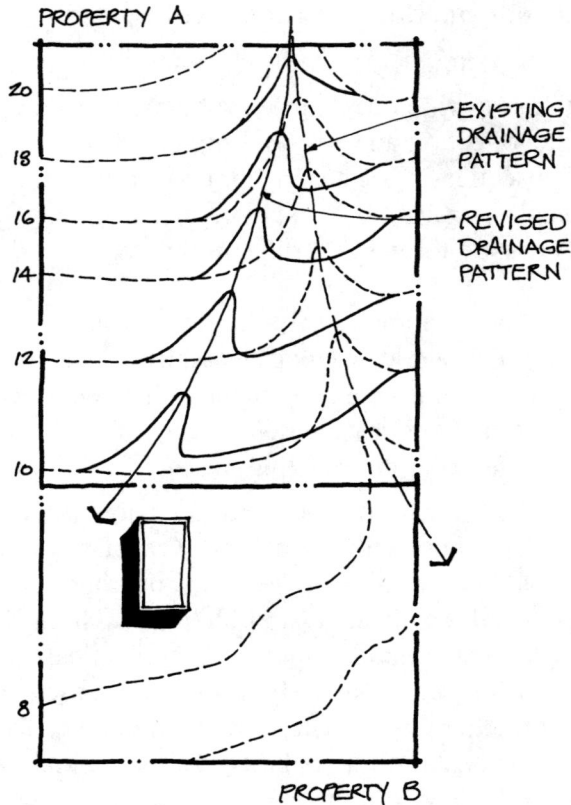

ALTERATION OF DRAINAGE PATTERN

Obviously, site analysis and design requires an understanding of the liabilities that may be incurred by altering drainage patterns, as well as the means to properly effect collection and disposal of surface waters.

Light, Air, and View Rights

An owner may require a light and air easement to prevent the construction of an adjacent building along a common property line, to insure the continued availability of natural light and fresh air to his building. Such an easement is generally obtained through negotiation. Because it often diminishes the use and value of the adjacent owner's property, a payment of money is generally required for granting such an easement. For example, a hotel owner may pay a substantial sum of money to limit the height of a proposed adjacent development, in order to preserve a view to the ocean from the hotel's rooms.

An *air right* grants a different owner the right to construct a structure in the air space above an existing building. Unless the structure is fully supported by adjacent structures, the right to

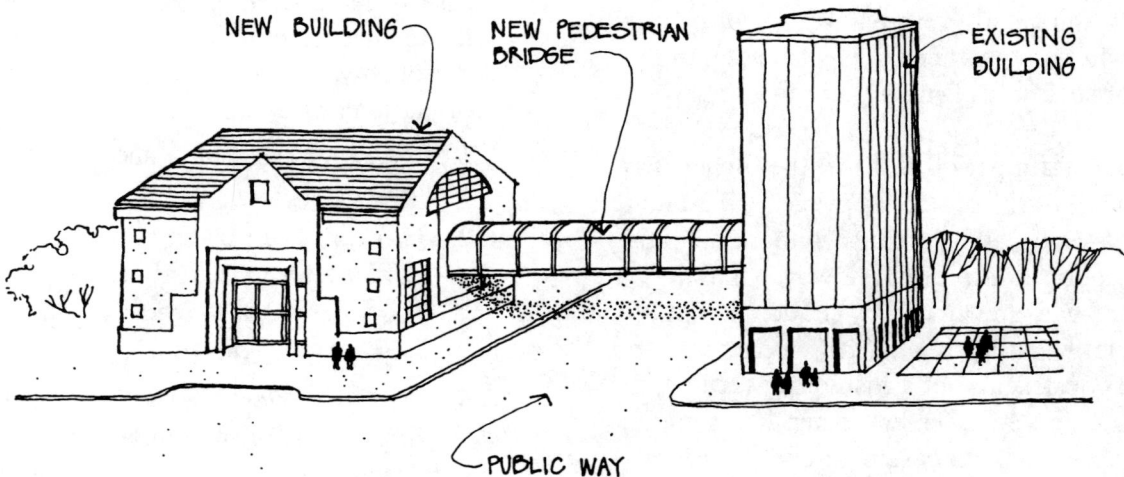

AIR RIGHTS OVER PUBLIC PROPERTY

construct foundations and provide access must be granted in the agreement. Because of the relatively high cost of such structures, air rights are usually restricted to areas where land values are very high.

Restrictive Covenants

Usually, one who conveys ownership of land to another person has no interest in how that land is used after the transfer of ownership takes place. But suppose the owner lives in the vicinity of the land transferred? More importantly, what if the owner is a developer who is selling lots or houses to a large number of buyers? The grantor may want to control land use after the title is transferred in order to assure buyers that the land will retain its character and value.

To accomplish these objectives, the developer can obtain express promises relating to the use of the land, called *restrictive covenants*. These restrictions, usually included in deeds of transfer, are binding on all future buyers, and enforcement rights are provided should any buyer violate them.

We have discussed numerous public and private land use controls affecting the potential development schemes of parcels of land. Each may significantly influence the design of sites and the arrangement of buildings and other improvements. To fully utilize the land under consideration, the designer must consider the implications of these controls in the analysis of a site in order to maximize its potential for development.

SITE DEVELOPMENT COSTS

Introduction

An architect usually strives to produce the best architectural solution for the least amount of money. Consequently, decisions made during the design phase must consider not only the project's aesthetic qualities, but also its initial expense and maintenance cost throughout its useful life. To be successful, the architect must evaluate alternative site development schemes, select an optimum solution, estimate various options, and compare costs and their impact on the total budget of a project. He or she must be familiar with materials, systems, and construction operations.

For example, an architect must be able to analyze slope and soil conditions and evaluate alternative building sites in order to recommend an optimum location for structures that minimizes the cost of earthwork, footings, and soil preparation for landscaping. Or the architect may have to review the costs of various methods of surfacing or illuminating roads and walkways and select the most cost-effective system available. This does not imply that the least expensive is always the preferred solution. The selection of site materials and systems must result in the appropriate solution that serves the owner's functional needs in the most economical manner. Consequently, the architect must evaluate and prioritize materials and systems and select those that are most responsive to the specific needs of the program.

Scope, Quality, and Cost

The design of a site, no matter how large or small, is affected by the scope of its development, the quality of the work, and the limitations imposed by the project's budget. An owner may establish the scope of the work based on anticipated needs, express a preference for a certain quality of materials and workmanship, and establish a total budget for the project. In such cases, the architect must retain the right to determine the affordable level of quality in order to assure the project's successful completion within the established budget.

BUILDING
LESS EARTHWORK
HIGHER BUILDING COST

SITE DEVELOPMENT
MOST EARTHWORK
HIGHEST COSTS

BUILDING
LESS EARTHWORK
LOWER BUILDING COST

SITE DEVELOPMENT
SOME EARTHWORK
LOWER COSTS

BUILDING
MORE EARTHWORK
HIGHER BUILDING COST

SITE DEVELOPMENT
MORE EARTHWORK
HIGHER COST

LEVEL   GENTLE SLOPE   STEEP SLOPE

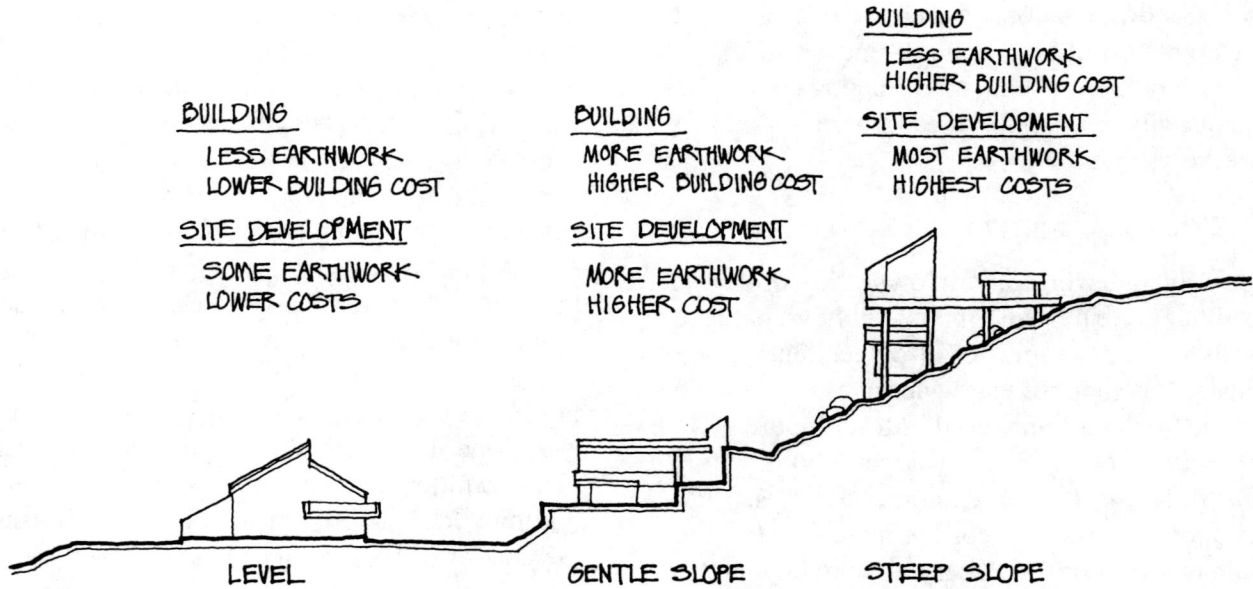

COMPARATIVE BUILDING AND SITE DEVELOPMENT COSTS

For example, if the cost estimate exceeds the budget because the owner wishes to use brick paving, and he or she is unwilling to reduce the quantity or increase the amount budgeted for paved walkways, the architect must insist on a reduction in quality of the paving material. He or she may suggest a compromise solution which utilizes a combination of paving brick and concrete to retain some of the desired quality and appearance without exceeding the budgeted cost.

Under no circumstances should all three aspects—scope, quality, and cost—be fixed: one or two must always remain variable. For example, if both scope and quality are rigidly defined in the program, then the budget must remain flexible. If the estimated costs exceed the budget, the architect must evaluate both scope and quality and recommend viable alternatives for the owner's consideration. This may imply, in the case of paved walks, for example,

either reducing the total area of paving or selecting a less costly material, or both.

Although the initial installation cost is more apparent during the development of a project, maintenance costs may have a greater impact on the total cost over the life of the project. This requires estimating the initial cost and the cost to maintain or replace the material or system over a period of time, and comparing one system against another. An example of this is the selection of a more expensive material such as brick, requiring no finish, instead of concrete, which may have to be painted initially as well as periodically.

Because construction costs are always changing and vary considerably from one part of the country to another, it is unlikely that exam candidates will be required to know specific costs related to site work, or to prepare cost estimates for a given scope of site development work.

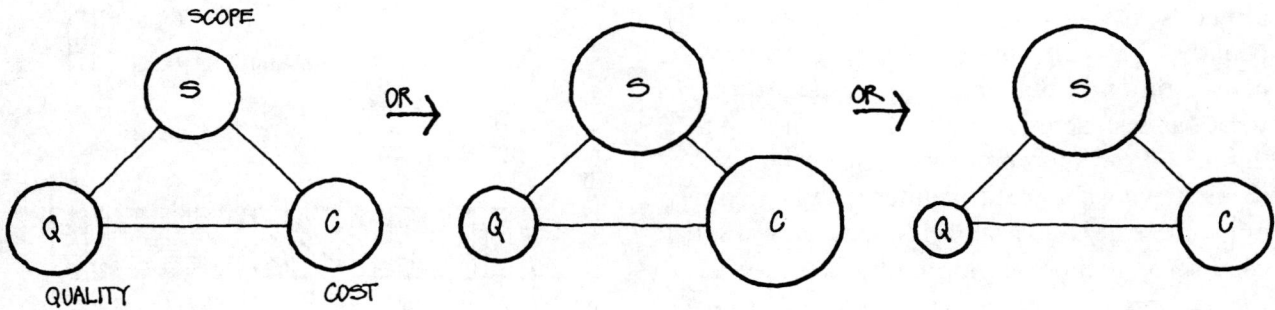

SCOPE , QUALITY , AND COST

Nevertheless, a candidate may have to demonstrate the ability to compare relative costs of various materials and systems, as well as their long range cost benefits. Aspects of site work will, most likely, include some of the following:

1. Demolition of existing buildings, site improvements, and natural features such as trees, rock outcroppings, etc., to make room for new developments

2. Earthwork, including cut, fill, compaction, etc., to prepare a site for buildings, roads, and other improvements

3. Foundations, considering the effect of soil conditions and topography on the cost of footings

4. Utilities, including the installation of new services as well as the extension of existing lines to provide the necessary services to a site

5. Paved roads and walkways

6. Landscaping

7. Lighting to illuminate roads, walks, paved areas, parking areas, and recreational areas

8. Site furniture, including benches or other seating, plant containers, and sports equipment

Initial Costs

Site construction costs are influenced by the cost of labor and materials, the efficiency of the contractor in managing the construction project, and the technology utilized to accomplish the work. Two important factors that affect site development costs are the characteristics of the site and the design of its development.

Unfavorable site conditions, such as a high water table, problems of access, poor soils, or steep slopes may result in excessive site development costs.

Careful attention in relating structures to topography, existing utilities, and vegetation, and the selection of materials and finishes, may exert an even greater impact on site development costs. Consequently, the architect must be aware of the comparative cost of the various elements comprising site work in order to select the optimum materials or systems, compatible with the overall project quality and budget.

Extensive earthwork, paving, storm drainage, or utility extensions increase construction costs, often requiring the deletion of landscaping or other site amenities so as not to exceed the project's total development budget. Consequently, if economical design were the principal consideration in planning a proposed project, what basic standards would the designer apply in selecting an appropriate site? He or she would recommend a site that is neither steeply sloping nor completely flat; a site with good natural soil, capable of supporting normal building loads, free of organic material, uncompacted fill, or rock within a few feet of the surface; and a site that has favorable natural drainage. The site would have a regular geometric shape and be served by the required utilities located close to its property lines.

In making preliminary recommendations, the architect would suggest a compact grouping of buildings so that connecting roads and walks could be kept to a minimum. Connecting elements would be arranged in a regular pattern that avoids useless leftover spaces, and have simple geometric shapes. Common areas would be compact, regularly-shaped, and concentrated in a central location, rather than spread out over the site. Landscaping would be limited to grass, in order to control dust; paved surfaces would be asphaltic concrete for roads and walks; and there would be very few, if any, retaining walls, raised planters, seating units, or other site amenities.

Fortunately, however, many owners do not insist on a "bare bones" project budget. Rather, they encourage the architect to incorporate site improvements which enhance the environment, and to select materials and systems which create an aesthetically pleasing, low-maintenance, well-balanced design.

Long Range Costs

Evaluating alternative materials and systems must include a comparison of maintenance costs as well as initial installation costs, because the

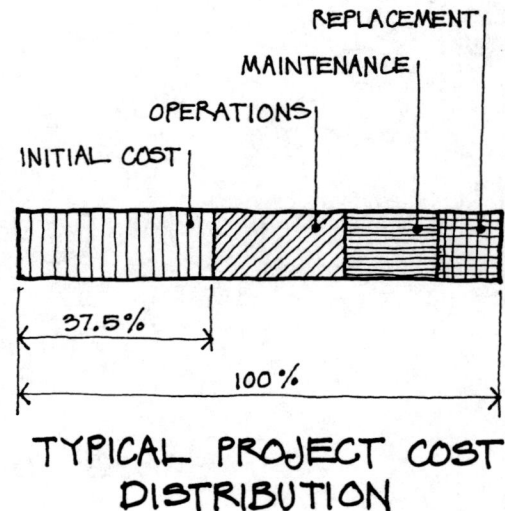

TYPICAL PROJECT COST DISTRIBUTION

latter represent only a fraction of the entire cost of a project over the period of its useful life. Value analysis is a systematic method of obtaining optimum value for every dollar spent, considering all project expenditures, including construction, maintenance, operation, and replacement. Although value analysis is not commonly applied to site work, the designer attempts to employ similar principles in order to obtain maximum value for the client's money, both initially and over the life of the project. The performance of a material or system over a period of years may be as important to an owner as its initial cost. For example, a durable paving material may be more appropriate where heavy use is anticipated than less expensive paving, which must be resurfaced at more frequent intervals. Heavy duty paving for roads may cost 50 percent more than light duty asphalt paving; however, if it lasts twice as long before requiring replacement, it will be 25 percent more cost effective for each life cycle of the material.

Since construction costs continually escalate, it is unlikely that candidates will be required to know unit costs for specific elements of site work. More likely, the exam will concentrate on comparative costs, cost benefits, and the

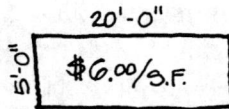

LIGHT DUTY

$6.00/S.F.
20'-0"
5'-0"

INITIAL COST = 100 S.F. × $6.00 = $600⁰⁰
LIFE CYCLE = 5 YEARS
ANNUAL REPLACEMENT COST = $600 ÷ 5 = $120⁰⁰

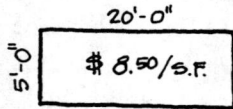

HEAVY DUTY

$8.50/S.F.
20'-0"
5'-0"

INITIAL COST = 100 S.F. × $8.50 = $850⁰⁰
LIFE CYCLE = 10 YEARS
ANNUAL REPLACEMENT COST = $850 ÷ 10 = $85⁰⁰

VALUE ANALYSIS

candidate's ability to select the most appropriate material or system for a given set of circumstances.

For example, the installation of a poured concrete paving slab over buried utility lines which require frequent maintenance is less cost effective than placing these utilities in a concrete trench with removable concrete covers, even though the latter is considerably more expensive initially.

Cost Control

To keep site development costs to a minimum, the designer must consider various alternatives in developing a design concept. The major elements comprising site development work and means to control their economic impact on projects are listed below. Cost effective design utilizes a logical approach to siting buildings and other improvements, resulting in an efficient physical organization of the site. The following examples illustrate that point.

1. Locate buildings along gently sloping terrain; avoid steeply sloping land. Why? To minimize costs of grading and excavation for building footings, retaining walls, roads, and utility lines. For example, locate a development of attached multi-family dwellings in a cluster along gentle slopes, somewhat above the base of the hill, rather than near its steeply sloping crest, in order to avoid excessive earthwork, unconventional building footings, and steeply sloping utility runs, and to provide ease of access.

2. Locate buildings where positive natural drainage exists. Why? To avoid redirecting surface flow by excessive regrading of the land and/or installing expensive storm drainage systems to dispose of storm water. For example, don't locate buildings and other site improvements at the base of a natural drainage basin, where rapid runoff of rain water requires costly methods of interception to avoid flooding.

3. Arrange vehicular circulation systems on the site to follow contours rather than to oppose them. Why? To minimize the cost of earthwork, construction of banks, berms, and retaining walls. For example, a two-mile long road following the existing contours will require considerably less excavation and associated sitework than a half-mile road up one side of the mountain and down the other, even though the former requires four times the amount of surface preparation and paving.

4. Locate paved parking lots on relatively level ground, rather than sloping ground. Why? To avoid excessive reshaping of the land, terracing, steps, and ramps to connect the various levels, and complex storm drainage systems to avoid standing water on paved surfaces. For example, a parking lot required to accommodate a large number of cars

located on a sloping site must be constructed in small, terraced segments connected by ramps for vehicles and steps for pedestrians. This requires more paving, curbs, and gutters, complex drainage devices, and more site lighting, all contributing to increased development costs.

5. Locate buildings so they relate to new and existing utility systems. Why? To minimize the length of connecting runs of utility lines. For example, a close physical relationship between utilities and buildings minimizes excavation, trenching, and service runs. Amenities requiring few or no utilities, such as tennis courts, may be located further from utilities.

6. Locate site improvements to utilize existing vegetation. Why? To avoid the removal of trees and plants and their replacement with costly new landscaping. For example, take advantage of the shading characteristics of mature deciduous trees by placing buildings north of them, rather than demolishing and removing them, grubbing, and grading the vacated area and preparing the site for new construction requiring new trees.

7. Avoid locating improvements over rock, organic soil, or areas of high water table. Why? To minimize costly excavation and foundation problems. For example, locating buildings where granite or limestone occurs close to the surface requires blasting prior to normal excavation, thereby increasing foundation costs considerably.

8. Coordinate the location of new with existing facilities including buildings, roads, walks, and other improvements in the development of a site design concept. Why? To preclude unnecessary and costly demolition and replacement of existing improvements. For example, try to utilize existing roads,

utilities, etc., in the design of a new or expanded site development. Should demolition costs be included in the project development budget? Yes!

9. Select appropriate finish materials for site improvements, including paving of roads, walks, plazas, play courts, etc., retaining walls, planter walls, seating and other site furniture. Why? To provide finished surfaces that best serve the specific function of the site and are in balance with the overall level of quality for the particular project, building type, and its anticipated life. For example, select broom-finished concrete in preference to slate or granite for campus walks, malls, and steps, to provide a useful, durable, easily maintained finish, appropriately priced for most construction budgets of educational facilities. (Note: Granite steps at the State Capitol, however, might very well be preferred.)

10. Select indigenous plant material for landscaping. Why? To minimize maintenance, irrigation, and replacement costs. For example, drought-tolerant plants (xerophytes) are preferred over water-loving plants (hydrophytes) in arid climates, where water is scarce and irrigation costly. Plants unaccustomed to a hostile climate will inevitably have to be replaced more frequently than indigenous species.

11. Select site lighting systems in consideration of capital cost, energy cost, and replacement lamp and labor cost. Why? To effect cost savings in the installation, operation, and maintenance of site lighting systems. For example, select high pressure sodium or metal halide lamps in preference to incandescent lamps for parking lot lighting, for lower energy consumption, ease of maintenance, and longer life.

LESSON 5 QUIZ

1. A residential use is permitted in which of the following zones?

   I. Residential zone

   II. Commercial zone

   III. Industrial zone

   A. I only

   B. I and II

   C. I and III

   D. I, II, and III

2. The purpose of restricting the number of dwelling units that may be built on a given plot of land is to control

   A. population.

   B. growth.

   C. density.

   D. cost.

3. Which of the following may be prescribed by zoning ordinances in order to maintain a certain amount of sky exposure?

   I. Maximum number of stories

   II. Maximum height

   III. Setback

   A. I and II

   B. II and III

   C. I and III

   D. I, II, and III

4. Floor area ratio (FAR) is the ratio of the floor area of a building to the

   A. total area of the site.

   B. buildable area of the site.

   C. maximum land coverage.

   D. area used for circulation.

5. The right to traverse a neighbor's land to gain access to one's own is known as

   I. an easement.

   II. a right-of-way.

   III. eminent domain.

   IV. condemnation.

   A. I and III

   B. I and II

   C. II and III

   D. I, II, III, and IV

6. To develop a shopping center in a district zoned for multiple dwellings, an owner would, most likely, apply for a(n)

   A. variance.

   B. conditional use permit.

   C. easement.

   D. code exemption.

7. The overall density of a cluster development for housing

   A. is the same as that for a conventional development.

   B. may be increased in exchange for certain contributions by the developer.

   C. requires the acquisition of additional land for the development of common recreation areas.

   D. varies depending on local needs for such facilities.

8. The main purpose of an environmental impact statement is to

   A. predict the consequences of a proposed action on the environment of an area.

   B. provide federal control over all types of physical development.

   C. assess the economic feasibility of a proposed project.

   D. identify environmental conditions endangering the proposed development.

9. Incentive zoning benefits

   A. the city.   C. the tenants.

   B. the owner.   D. all of the above.

10. Foundation costs are LEAST likely to be affected by

    A. soil conditions.

    B. topography.

    C. water table.

    D. vegetation.

PERCEPTUAL FACTORS

HISTORICAL BACKGROUND

During the Paleolithic Era, site planning did not exist. People lived on the land and took their shelter and food from the land, but they made no attempt to alter their surroundings. If site design is defined as the conscious rearrangement of the environment for human use, then site design began when man abandoned the cave. This dramatic development dates from the Neolithic Era, when people learned to cultivate crops, domesticate animals, and build shelters.

The earliest settlers, who tilled the soil and tended their herds, gathered in friendly groups to construct villages in which the inhabitants enjoyed the advantages of mutual shelter and safety. These ancient settlements inspired two basic forms that influenced all subsequent planning: the *rectilinear layout and the circular layout.*

The rectilinear grid arrangement responded to the need for planning, dividing, and measuring land, which were essentially agricultural requirements. Although the grid was inspired by the parallel lines of the plowed fields, it influenced the geometry of the village and the construction of shelters, as well. The circular layout was probably originated by herdsmen who required protection and control of their animals. It was an ideal form, because it enclosed a maximum of land area with a minimum length of

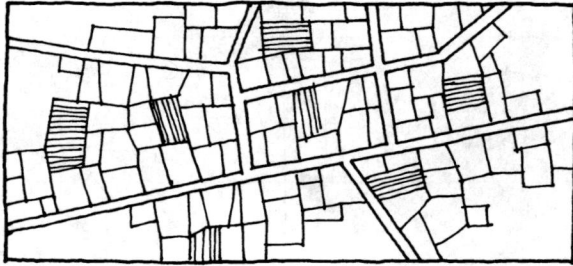

RECTILINEAR LAND LAYOUT

enclosure. Its major use, however, was in the layout of early fortified towns, whose protective walls were often encircling.

There were numerous patterns of urban growth, but whether rectilinear, circular, or irregular, expansion was generally outward, in concentric layers of development. With the advance of civilization, irregular and geometric patterns were superimposed one upon another, producing the variety of forms we see today. The historic patterns of site design, therefore, were frequently the result of continuous remodeling through the ages.

CIRCULAR LAND LAYOUT

The Ancient Civilizations

The earliest civilizations developed along the fertile valleys of the Nile, Tigris-Euphrates, and Indus Rivers, where food, water, and transportation were available. Societies evolved into kingdoms, which were governed by powerful monarchs who guided all urban development.

In Egypt, Mesopotamia, and the Indus Valley, permanent towns were planned and constructed in a generally rectilinear pattern. The simple dwellings of these towns housed the slave-workers who built the monumental temples, palaces, and tombs of the royal rulers.

Around 2100 BC, the earliest building regulations were embodied in the codes of the Babylonian King Hammurabi. According to decree, irresponsible builders received harsh punishment, sometimes even death, if their work caused injury to others. The history of these early civilizations is a chronicle of wars, revolts, and conquests, reflecting the plight of the masses, as well as the insatiable appetites of their rulers for power.

The Classic Civilizations

A more enlightened society developed in the early communities around the Aegean Sea, and it is in Ancient Greece that we find the principles and models upon which Western civilization was based. The Greek sense of the finite demanded that all things, including the design of communities, should be of a limited size in order to be workable and comprehensible. Few towns had a population greater than 10,000, since that was the practical limit for producing food, distributing water, and transporting products in any one area. When a town exceeded its largest practical size, development was stopped and a new town was established at another nearby site.

The earliest Greek cities were irregular in form, with a maze of narrow streets rambling over the rugged topography. In the latter part of the fifth century, however, a planner named Hippodamus originated a number of formal city planning theories, among which was the gridiron street pattern. Thereafter, the grid arrangement was superimposed on the rocky hillside sites of most cities, disregarding the natural terrain and occasionally producing precipitous streets that required steps.

The *agora*, or market place, was the center of business and political life, as well as the geographical center of town. The major city streets terminated at the agora, rather than crossing it, because the generous central space was intended for pedestrian circulation and assembly. Dwellings were arranged around south-facing courtyards and grouped in geometric blocks, while sports and recreation were concentrated in the gymnasium, stadium, and theater. Shrines, temples, and other public structures were located near the agora, but adjacent to their own open spaces.

While Greek planning had a sense of the finite, the Roman approach was to overcome the problems of growth through technological achievement. Thus, aggressive city builders and skilled engineers developed sophisticated water supply, drainage, and circulation systems that enabled Roman cities to expand and flourish. As Rome continued its unrestrained growth, however, the quality of life diminished. In time, the cities grew congested, the rulers moved to their spacious villas in the country, and Rome suffered decay, deterioration, and decline.

Medieval Developments

The deterioration of the Roman Empire led to the Dark Ages, when cities declined in importance and the urban population reverted to a rural life and an agricultural economy. Towns of the Middle Ages were small; their practical limits were determined largely by the ability of the surrounding land to support the dependent population. Their labyrinthine forms were based on military defense; they were often located on abrupt terrain and surrounded by heavily fortified walls, and irregular streets, radiating from the town centers, twisted their way to the city gates.

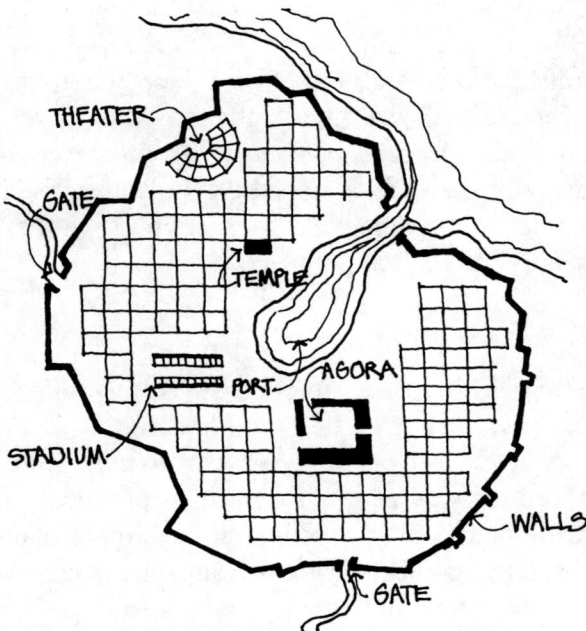

TYPICAL GREEK TOWN

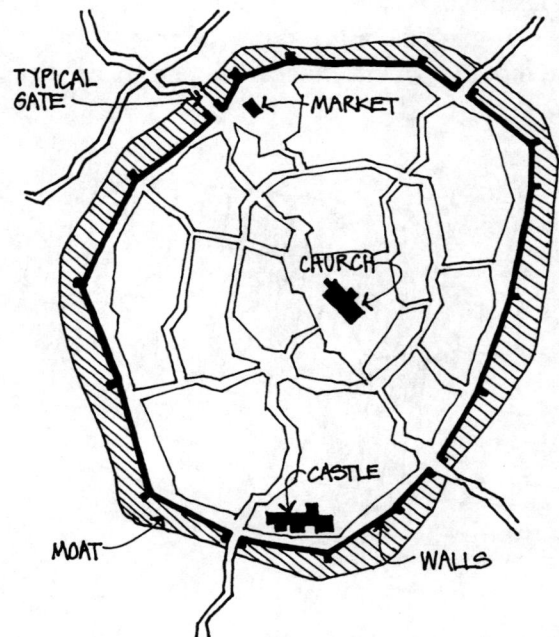

TYPICAL MEDIEVAL TOWN

The town was dominated by the centrally-located church, before which a spacious plaza served as the site for commerce, social interchange, and entertainment. When the population of a Medieval town was small, space was ample and life was tolerable. As the population grew, however, new buildings filled in the open spaces, congestion led to lack of hygiene, and the tragic results were disease, decay, and pestilence.

Renaissance Developments

The invention of gunpowder in the 15th century marked the end of the walled town, and coincidentally, the beginning of the Renaissance. Urban planning during this period became more intellectual, more formal, and particularly, more monumental. Every form had its centerline and every space its axis. Cities were patterned after classic models, and the arts became the tools by which wealthy merchants, churchmen, and kings produced their enduring monuments. The Renaissance introduced spacious urban plazas, broad boulevards, and vast, formal gardens, all of which represented an attempt to provide expansive open space within the city.

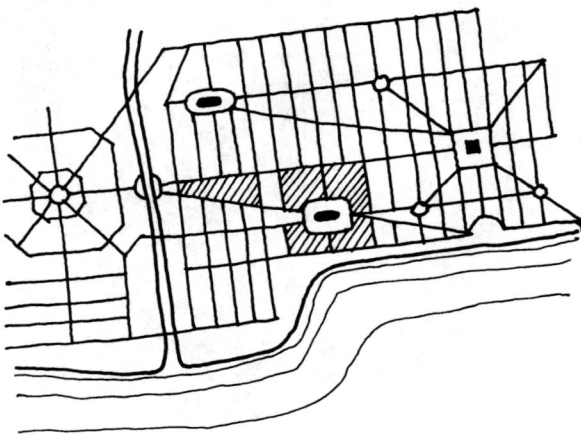

TYPICAL BAROQUE LAYOUT

In later years, the influence of French urban design concepts could be seen in Vienna, St. Petersburg, Athens, Mexico City, and in the first significant city planning effort in this country, Washington, D.C. It was Pierre Charles L'Enfant, with his background in the baroque idiom of Paris and inspired by the American spirit, who developed a grandly classic plan for the new capital. L'Enfant's geometric plan, which had diagonal and radial streets superimposed on a typical gridiron layout, appealed to the aristocratic tastes of Washington and Jefferson, and it was adopted for implementation in 1791.

Thomas Jefferson's approach to achieving urban space and fresh air was embodied in his proposal to limit development to only half of the site. His favorite city plan resembled a checkerboard grid pattern in which every other block was a park. Parts of Jeffersonville, Indiana, were actually laid out in this fashion at the beginning of the 19th century.

The bold rebuilding of Paris by Baron Haussmann, during the mid-19th century, culminated this persistent effort to achieve urban open space. It was one of the most extensive urban transformations ever undertaken. Haussmann created broad new boulevards, tree-lined streets, and spacious gardens, resulting in the beauty we see today in Paris.

The Machine Age

The Machine Age, which began in the 18th century, developed into the Industrial Revolution in the 19th century, and it profoundly influenced the way people lived. Mass production made it possible for people to have more manufactured goods, and the advancements in transportation and communication forever changed the urban culture. As factories proliferated, the number of workers drawn to urban centers increased, as well. The economic shift from farm to factory was accompanied by a soaring expansion in population density. The speed of this dramatic change took its toll on the urban environment in the form of congestion, pollution, and painfully unhealthy living conditions.

A number of European theorists developed schemes for model worker towns that attempted to solve these urban problems. New communities, intended exclusively for workers, were designed to be totally self-sufficient, complete with recreational and educational facilities. Worker communities were located near major industrial plants, such as the English towns of Port Sunlight, built by Lever Brothers, and Bourneville, a garden community near Birmingham, that was built by the Cadbury Chocolate Company. In this country, Gary, Indiana, was planned by the U.S. Steel Corporation in accordance with its precise requirements, and named for its chairman. Pullman, Illinois, too, was built as a permanent town for workers employed in the factory that manufactured Pullman sleeping cars. The company town concept in this country culminated with the development of Kohler, Wisconsin, built by the plumbing fixture manufacturer.

The Machine Age influenced a number of European planners, most notably the French architect Tony Garnier, who designed a hypothetical industrial town called La Cite Industrielle. His suggestion of land uses anticipated modern zoning laws. Some years later, the Dutch architect J.J.P. Oud designed a small colony for workers near Rotterdam which embodied Garnier's theories. The Spanish engineer Soria y Mata proposed a Linear City, in which the logic of linear utility lines became the model for the city layout. One such city was

actually built near Madrid, and this same idea served as inspiration for the planned linear city of Stalingrad. A similar concept was advanced by the Italian futurist Sant' Elia in his La Citta Nuova. This was a visionary design of an enormous metropolis, whose principal elements were based on circulation channels.

The Columbian Exposition of 1893 in Chicago, for which Daniel Burnham was the chief architect, launched a classic revival movement that became known as the *City Beautiful Era*. This movement inspired the construction of scores of civic structures across the country, all based on classic architectural concepts. Burnham proclaimed, "Make no little plans," and nearly every major American city obliged. The sudden awareness of the aesthetic qualities of communities and the life styles of their citizens evoked a new social awareness among urban planners. Out of their reaction to the deteriorating environment of the industrial communities emerged the idea of the *Garden City*.

Twentieth Century Developments

At the close of the 19th century, distinguished social critics such as Ruskin, Viollet-le-Duc, and Thoreau were joined by the Scottish planner Patrick Geddes, who spoke out for regional planning and thereby gave impetus to the Garden City concept. An influential theoretician, Ebenezer Howard, delineated the particulars of the Garden City in his book *Tomorrow*, published in England in 1898. Disturbed by the unhealthful conditions of cities, their haphazard growth, and their lack of aesthetic quality, Howard proposed the creation of a city on 1,000 acres with 30,000 inhabitants. A central core of public buildings was surrounded by commercial shops and dwellings, with industrial facilities located on the city's outskirts. All of this was encircled by a permanent green belt of about 5,000 acres.

Howard also proposed that the community own and derive all the benefits from the land, thereby discouraging speculation. Early in the 20th century, two Garden Cities, Letchworth and Welwyn, were developed with the intent that they be completely self-sufficient. Both cities, however, became dependent on nearby London for employment, and consequently they became satellite towns instead of true Garden Cities.

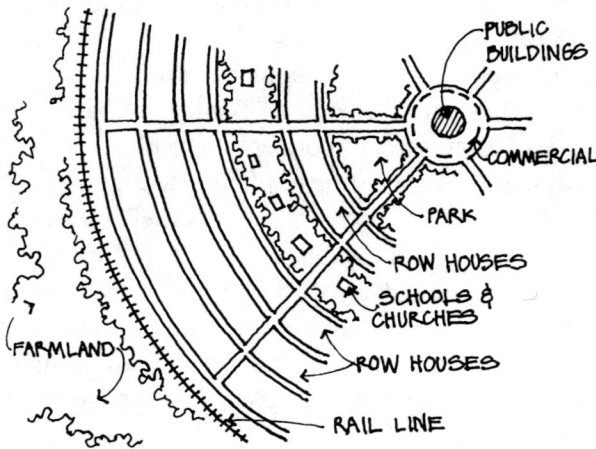

HOWARD'S GARDEN CITY SCHEMATIC

Inspired by the Garden City idea, the American planners Henry Wright and Clarence Stein developed, during the 1920s, a concept of organizing towns into cohesive neighborhoods. To minimize conflicts between automobiles and pedestrians, they conceived the *superblock*, which was an island of green, bordered by dwellings, with roads and parking placed at the periphery. These ideas were demonstrated by Wright and Stein in the planned community of Radburn, New Jersey.

The superblock of Radburn comprised an open green space of nearly 50 acres, within which there was no through traffic, since automobiles were relegated to the peripheral roads. Within the superblock, houses were grouped around cul-de-sac roads, which provided vehicular

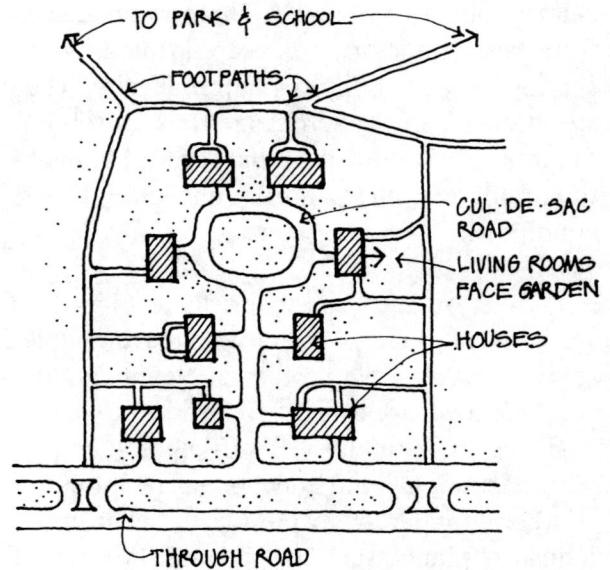

SITE PLANNING AT RADBURN

access to each unit. Pathways, located at the perimeter of each cul-de-sac house group, led to parks and schools. Dwellings were oriented with the garages facing the road, while the living areas were turned toward the gardens—the reverse of conventional siting at that time. Although never completed, Radburn was one of the most influential planning concepts ever developed for the modern residential community.

Several 20th century architects, including Le Corbusier and Frank Lloyd Wright, addressed the problems of urban planning and set forth their individual theories. In contrast to Le Corbusier's proposal for high density towers surrounded by open green spaces, Wright, in his plan for Broadacre City, proposed that each dwelling unit be located on an acre of land. If Wright's plan had been followed, it most likely would have produced an endless suburban sprawl.

During the Great Depression, in 1935, the federal government engaged in an experiment with "greenbelt" towns. Although the designs were inspired by the Garden City idea, they were not planned to be self-contained towns, but rather satellite dormitory villages for nearby cities. Each was surrounded by a belt of permanent open space and provided with a full complement of community facilities, including schools and commercial and recreation areas.

A significant example of this plan type was Greenbelt, Maryland, located about thirty minutes from Washington, D.C. The development included about 1,000 living units located on about 12 percent of the 2,000 acre site. Each superblock contained approximately 120 dwellings with interior play areas. With the addition of underpasses, conflicts between pedestrian circulation and automobiles were nearly totally avoided. The commercial and community center at the core of the plan reduced to a minimum the walking distance from the dwelling units.

This brief synopsis of site design history is intended as an overview of developments that have led us from the cave to the urban centers in which most of us presently live. It has been said that "we have need to learn from yesterday so as to prepare today for a better tomorrow." Designers must be aware of previous developments in order to gain a better understanding of the problems and opportunities facing today's urban communities. With this understanding, planners can more ably respond with vision, wisdom, and a sense of environmental responsibility.

SITE ANALYSIS

General Purpose

Site analysis is the process of investigating basic data that relates to a particular site, such as survey information, topographic data, geological information, zoning ordinances, existing character, microclimate, development patterns, social patterns, etc. The purpose of site analysis is to determine whether a parcel of land is suitable for a specific proposed use. It would be undesirable, for example, to situate a school adjacent to a major freeway. Similarly, a roadside cafe should not be located out of sight of the road, nor should a meat packing plant be placed upwind of dwelling units. All of these are examples of inappropriate uses for a given site, or perhaps inappropriate sites for a given use. In theory, almost any site will support almost any use; however, the ideal situation is one that most fully satisfies the project criteria, while requiring the least modification or compromise.

Every site is as unique as an individual person, and even as identical twins have distinct personalities, so too, adjacent parcels of land possess distinguishable characteristics. Part of this uniqueness is reflected in a site's equilibrium. The flow of surface water creates a discrete drainage pattern, plant and animal life constitute an ecological system, and human use conforms to a workable social structure. Site factors such as these are interrelated, and at any given moment they are in balance, even if they are in the process of change. The recognition of a site's character reveals the practical limits imposed on a planner, as well as the potential damage that may be inflicted by development.

All development implies change, and occasionally this change produces undesirable effects. Excavation, for example, may alter drainage patterns, grading may cause erosion, and the construction of new facilities may destroy plants, pollute the air, or create traffic congestion. Environmental changes, therefore, are an inevitable result of the development process.

Relevant Data

The relevant site data which must be gathered and analyzed comprise those factors that determine the suitability of a site for its proposed use. Although no single set of factors applies to every situation, the following list includes data that is relevant in most circumstances. Some of these factors are described in greater detail in the previous lessons.

Climate

Every site is affected by regional climate patterns, as well as the microclimate that applies to a small area. Climate is related to topography, slope orientation, vegetation, and the presence of water, and it is important because it bears directly on human comfort. Climatological data may be obtained from the National Weather Service, through talks with local inhabitants, and by personal observation of weathered structures and existing plant material. The following features may be analyzed:

1. Temperature averages and extremes
2. Precipitation averages and extremes
3. Snowfall averages and extremes
4. Wind intensity and directions
5. Humidity patterns
6. Solar angles
7. Days of sunlight
8. Frost data

Topography

Topography is the form of a site's surface features, and it is a factor that strongly influences land development. The gradient of roads, disposition of structures, and visual aspects of a site are all influenced by the character of the landform. Topographic data is available from the U.S. Geological Survey aerial photographs, or

on-site surveys, and the features that may be analyzed are:

1. Elevations
2. Slope amount and direction
3. Unique landforms
4. Natural drainage patterns

Soils

Knowledge of the soil conditions on a site is important to determine the soil's capacity to support buildings and roads, as well as its ability to sustain plant material. Soils data is obtained from the U.S. Department of Agriculture Soil Conservation Service, test borings, visual inspection, and the experience of neighboring developers. The following features may be analyzed:

1. Soil types
2. Moisture content
3. Depth of organic topsoil
4. Depth to water table
5. Depth to bedrock
6. Drainage characteristics
7. Susceptibility to compaction
8. Soil fertility
9. Rock outcroppings

Hydrology

Hydrology refers to the occurrence, movement, and quality of water on a site. Surface water and drainage patterns affect vegetation, climate, and potential development, and this data is available from the U.S. Geological Survey, local hydrological studies, and on-site inspections. Hydrological considerations include:

1. The form of surface water (streams, lakes, etc.)
2. Drainage patterns
3. Runoff rates

4. Subsurface water characteristics

5. Aquifer (water-bearing) zones

Vegetation

Plant types and patterns represent a major site resource, and they contribute significantly to the unique character of an area. Native landscaping is closely related to climate, hydrology, and topography, and it often determines the form of development.

Data on vegetation is available from U.S. Geological Survey maps, aerial photos, and on-site observation. The factors that may be analyzed are:

1. Types and extent of vegetation

2. Density of vegetation

3. Heights of vegetation

4. Health of vegetation

✓Existing Land Use

As a site is developed, man-made features become more important than natural features. Structures, circulation systems, and activity patterns must be considered. Such data is obtained from land use maps, state highway maps, historical preservation societies, and personal inspections, and some of the factors that may be analyzed are:

1. Existing roads and paths

2. Existing utility lines

3. Existing air and rail facilities

4. Type and number of structures

5. Uses of open space

6. Human behavior patterns

7. Historical sites, structures, and trails

Sensory Qualities

The sensory qualities of a site are those intangible elements that affect people through the senses of sight, smell, touch, and hearing. The uniqueness of a site may be its view or its geometry, the smell of wildflowers or of the ocean, the feel of heat or of wind, or the sound of traffic, church bells, or singing birds. The perception of sensory qualities is as important to site analysis as any other relevant factor, and pertinent data of this sort is almost always obtained through first-hand, on-site observation. The features to be analyzed may include:

1. Scenic vistas

2. Spatial illusions

3. Quality of light

4. Characteristic smells

5. Characteristic sounds (noises, echoes, etc.)

6. Sensation of natural forces

7. Perception of textures

Natural Hazards

There are several natural elements that are potentially hazardous to certain types of development, and others, such as earthquake faults, that may restrict almost all construction. Information on hazards is generally available from a variety of government agencies, local inhabitants, and sometimes (unfortunately) through personal experience. Analysis may include:

1. Earthquake fault zones

2. Hurricane zones

3. Tornado zones

4. Flood plains

5. Tidal inundation areas

6. Wet zones (peat bogs, quicksand, etc.)

7. Areas of poisonous plants

8. Areas of poisonous snakes or reptiles

9. Areas of annoying insects

The actual site analysis begins when all of the pertinent information is collected. At this point, a base map is prepared showing legal boundaries, contours, roads, buildings, utilities, and other natural or man-made key features. The base map is used as a background on which various overlays are produced, generally one for each area of concern. For example, a soils overlay may classify soils by type and depth, with locations and logs of known test borings. A visual survey overlay may consist of personal notes and observations regarding scenic views or unsightly features in need of modification or removal. When environmental concerns are explored, the resulting overlay may serve as a checklist for an environmental impact assessment.

A map on which all the overlaid information has been superimposed is known as a site analysis map, an example of which is shown on the following page. This map indicates the degree to which a site is suitable for a proposed function. At this point, the planner may discover that compromises may be necessary. For example, a site that appears optimum for a shopping center, based on population growth studies, topography, suitable soil, costs, etc., may be located too far from freeway access. For the shopping center and freeway access to be closer together, one may be forced to accept a lower quality site. Therefore, when a site is judged to be suitable for a proposed use, it is almost always a matter of striking a balance between what is ideal and what is reasonably possible.

NATURAL LANDFORMS

Landform refers to the shape of the earth's surface, which may include everything from mountain ranges to furrows in a field. Landform is important because it affects the aesthetic character of an area, as well as one's perception of space. Level land, for example, unifies the landscape, while hilly land tends to divide it. Landform types also have a direct impact on the development with which they are visually compatible. The natural shape of land, therefore, affects how it is perceived, modified, and used.

Landform may be classified by character, steepness, geology, etc., but where visual, functional, and perceptual qualities are concerned, the most significant factor is form. The landscape is a continuous composition of varying earth forms that blend into and reinforce one another. Where a level stretch of desert ends, the concave slope of a mountain begins, but that precise point may be difficult to detect. Similarly, the merger of a slope and a valley may be nearly imperceptible. Following is a discussion of the most common landform configurations with some implications of their potential for site design.

Level Landforms

A level landform is any area that appears visually parallel to the horizon. Of course, there is no such thing as a perfectly level piece of land, because all ground has some amount of slope. Nevertheless, land perceived to be level is stable, static, and in equilibrium. Level land is comfortable because it requires little effort to stand, walk, or rest on a surface that is in balance with the earth's gravitational forces. For these same reasons, level areas are the most sought-after sites for buildings. In fact, when level sites are not available, they are often created by remodeling sloping terrain into flat pads.

Level landforms lack spatial definition; other than the horizon, no elements appear to enclose space. On the other hand, this openness permits extensive, uninterrupted views, which establishes a unifying force on the landscape. Level land induces a feeling of exposure; there is no protection against sun or wind, there is no

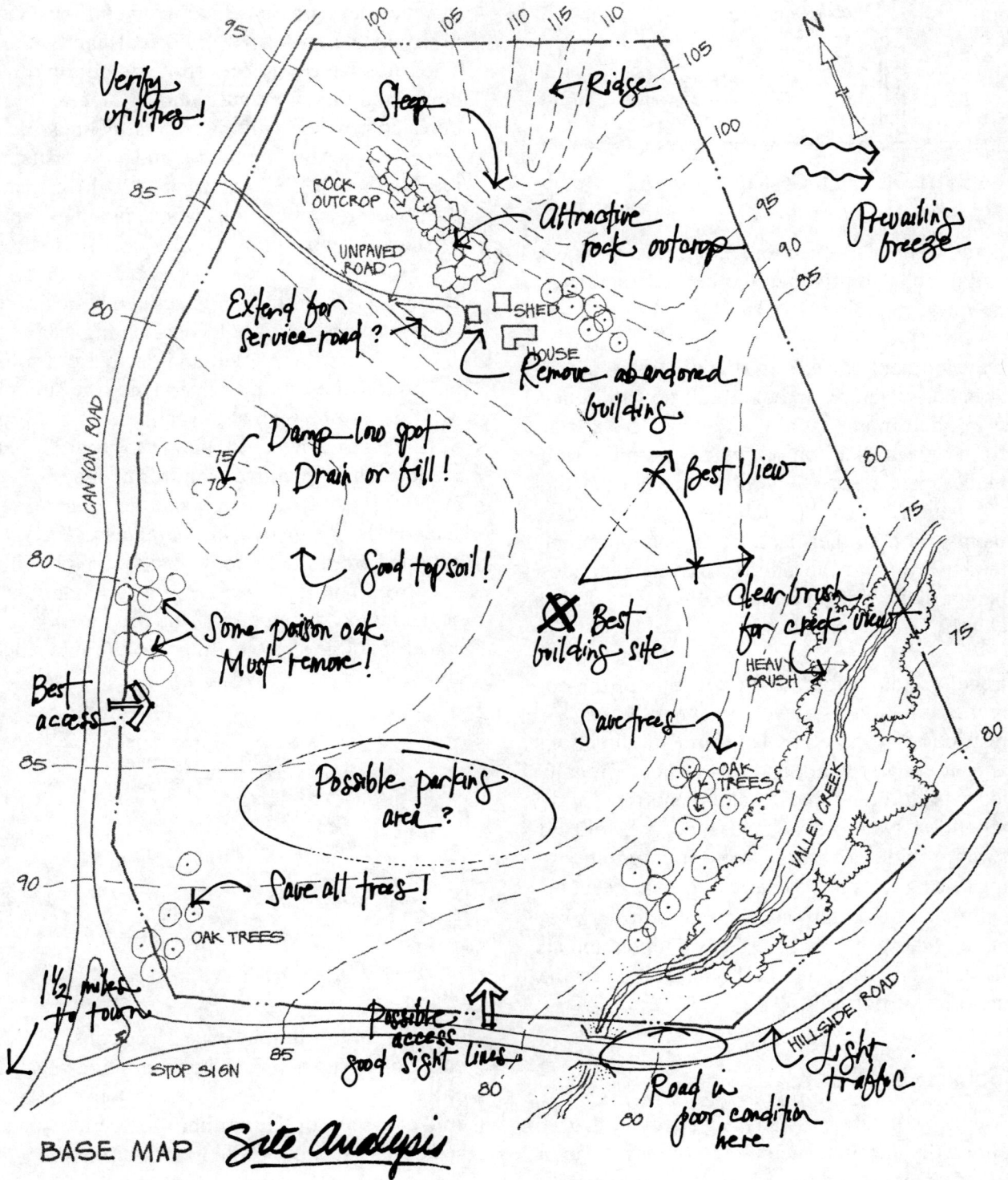

Verify utilities!

95   100   105   110   115   110

N

← Ridge

105

Steep

100

ROCK OUTCROP

95

Attractive rock outcrop

85

UNPAVED ROAD

90

85

Prevailing freeze

Extend for service road ?

SHED

HOUSE

Remove abandoned building

80

CANYON ROAD

Damp low spot Drain or fill !

75
70

Best View

80

75

80

Some poison oak Must remove !

Good topsoil !

Clear brush for creek views

75

Best building site

HEAVY BRUSH

Best access

85

Possible parking area ?

Save trees

OAK TREES

VALLEY CREEK

80

90

Save all trees !

OAK TREES

1½ miles to town

Possible access Good sight lines

HILLSIDE ROAD

Light traffic

85

STOP SIGN

80

Road in poor condition here

80

BASE MAP   **Site Analysis**

DOMINATING VERTICAL ELEMENTS

HARMONIOUS HORIZONTAL ELEMENTS

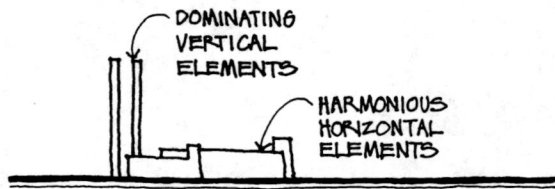

LEVEL LANDFORM

defense against objectionable noises or views, and there is no privacy. In other words, there is no place to hide.

Development of level land is relatively unrestricted; structures may be built upward, outward, and in almost any direction. Horizontal forms appear harmonious when set on the level landscape, because they reflect the horizon and emphasize the earth's stability. Vertical elements, on the other hand, attract attention and tend to dominate the landform. Even the modest height of a farm silo on the Midwestern plain is clearly visible for miles.

Level ground offers little indication of the correct orientation for development. With no restrictions imposed by landform, all directions appear equally valid, and this has occasionally led to multidirectional developments that sprawl repetitiously across the landscape. A level landform provides a neutral, sometimes uninteresting setting that can be characterized as peaceful, calm, and quiet, although in reality it may be none of these. Nevertheless, level topography is flexible, practical, and highly desirable for the majority of human uses.

Convex Landforms

A convex landform is any high ground that is outwardly curving, such as a knoll or mountain top. Compared with level landforms, convex forms are dynamic, exciting, and powerful, in

that they defy gravity. Historically, hilltops have always held a strong psychological significance. The tribe that occupied a strategic hilltop often controlled the entire surrounding area, and an elevated position still engenders feelings of respect. We "look up to" the judge who sits on a raised bench, and we hold in "high" esteem the government buildings whose broad expanse of steps we must climb.

Because high ground possesses symbolic importance, projects that are located on high ground are considered important and desirable. Building on top of a summit can enhance the summit itself, as can be seen at the Monastery of Mont S. Michel in France. In addition, projects that are located atop convex landforms may be highly prized for their remarkable orientation, such as the views over San Francisco Bay obtained from the famous hotels on Nob Hill. A convex landform can serve as a focal point or landmark, depending on its prominence. It can also define space by means of its side slope and summit.

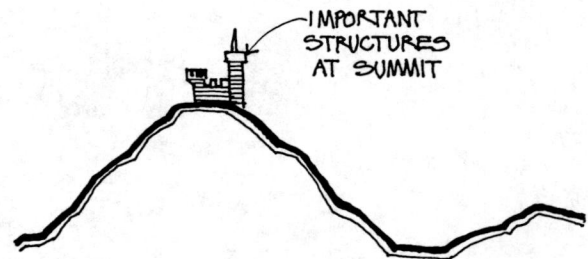

IMPORTANT STRUCTURES AT SUMMIT

CONVEX LANDFORM

Convex landforms may experience considerable variation in sun, wind, and precipitation because of their orientation. South-facing slopes are generally favored in the northern temperate zone because of their exposure to sun during the winter, as well as their protection from cold winter winds and precipitation that originate from the northwest. Both factors contribute to the energy efficiency of structures placed on south-facing slopes.

Concave Landforms

Concave landforms are depressed areas in the landscape that are depicted in plan by concentric and rising contour lines. They are low areas of land in which a variety of human activities take place. All concave landforms have a relatively high degree of spatial enclosure, since they are defined by surrounding slopes.

Concave forms are generally oriented with the focus toward the center, and they produce feelings of seclusion, privacy, and protection from the surrounding environment. They are confining, however; views are restricted, there is a lack of connection to outside spaces, and they are vulnerable from higher, surrounding slopes.

SECLUDED PRIVATE & PROTECTED

CONCAVE LANDFORM

Concave forms are suitable for activities requiring a sense of enclosure and an inward focus, such as the outdoor performances held in the natural landform known as the Hollywood Bowl in Los Angeles. Other, more extensive concave forms have served as sites for entire communities. Since concave landforms are generally protected by surrounding slopes, they are warmer and less windy than other landforms. However, precipitation that falls on a concave form drains to its bottom floor and may create perpetual dampness if not carried away. For this reason,

concave landforms make suitable sites for retention basins and even permanent lakes.

Ridge Landforms

A ridge is a convex landform that is linear. In other words, rather than a high point, it is a succession or line of high points along the ground. A ridge clearly defines space, since its form acts as a limiting wall. The linear or directional characteristic of a ridge is capable of leading the eye along its length; conversely, ridges provide numerous vantage points for wide-angle views.

Ridges are functionally desirable for circulation routes located on top or parallel to the ridge. Drainage is good, views are often spectacular, and related development can be located adjacent to the road, just off the ridge line. Ridges are suitable for all types of linear developments, but they are inappropriate for dispersed or sprawling layouts that require more extensive level areas.

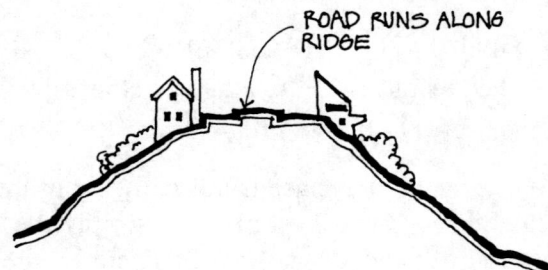

ROAD RUNS ALONG RIDGE

RIDGE LANDFORM

Valley Landforms

A valley is the reverse of a ridge; it is a concave low area that is linear and directional. Valleys, like ridges, are suitable for circulation routes, and historically, this has been one of their principal functions. Valley floors are often fertile areas, and many of the country's great agricultural regions are located in this type of landform. Because of their quality of spatial enclosure, valleys are also suitable for developing communities, which has often led to rivalry for choice land between developers and farmers.

Valleys are popular activity areas because they convey feelings of safety, refuge, and seclusion. They are also associated with stream beds, since they represent a topographic form that is suitable for natural drainage. Permanent development, therefore, should avoid the lowest levels of valley floors.

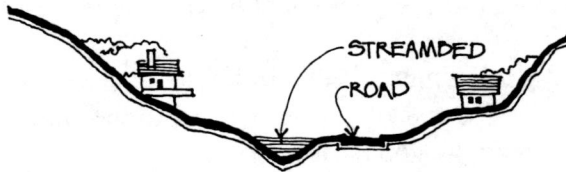

VALLEY LANDFORM

Perception of Space

The perception of space is influenced by three elements of landform:

1. The base area of the space
2. The steepness of the adjacent slopes
3. The visible horizon line

The base area of a space is normally the usable area and may be a level plane or simply the bottom of the adjacent slopes. Generally, the larger the base area, the larger the space is perceived to be. When one is standing in a valley, the adjacent slopes act as walls, and therefore, the steeper the slopes, the more enclosed the space will feel. When standing atop a ridge, on the other hand, one will have a wide-open feeling as the adjacent slopes fall away.

The visible horizon line is the boundary of one's perception of space, and is a function of its height and distance from the viewer. In a broad valley bounded by low hills, for example, space will feel far more expansive than in a deep

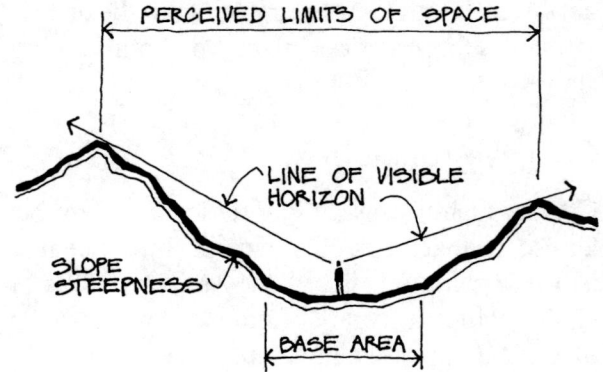

PERCEPTION OF SPACE

ravine formed by precipitous cliffs. On level land, the visible horizon line may be where the land meets the sky, in which case space is perceived as nearly infinite. All three variables, base area, slope steepness, and horizon line, interact to affect one's perception of space. Moreover, the degree of spatial enclosure can be manipulated by varying any one of the three elements. Perception can be manipulated as well, by sloping the base plane, which creates an unstable space, or varying the slopes from one side to another, which produces a spatial orientation away from the steep high ground and towards the shallower, lower side.

Regulating Activities

Landforms can be manipulated to control views, that is, reveal or conceal focal points, as well as regulate the direction, speed, and rhythm of movement. Views are directed toward open spaces, along lines of least resistance. Therefore, built-up landforms on both sides of the sight line create blinders that focus the view straight ahead. Objects placed on slopes are readily seen from opposite slopes or from below, because the slope acts as a background that arrests the vision. Objectionable elements can be screened out by landforms, such as the use of earth berms along parking lot edges to block the view of cars. The crest of natural

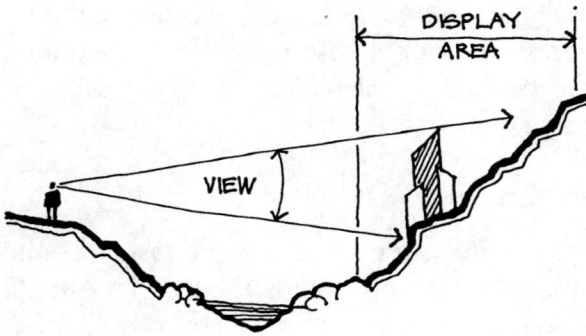

VIEWING OBJECTS ON SLOPES

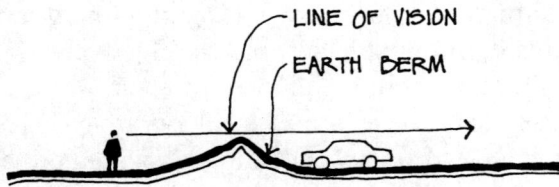

SCREENING VIEW WITH LANDFORM

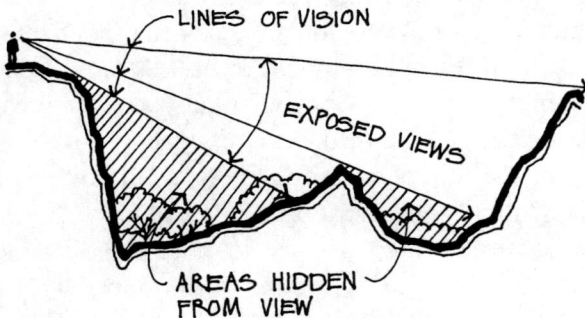

REGULATING VIEWS WITH LANDFORMS

slopes may conceal unpleasant objects, if these objects are located behind the slope or substantially below the slope from a viewer.

Patterns of movement may also be influenced by landforms, because circulation follows the path of least resistance. Level areas, of course, require less energy to traverse than irregular terrain, and

sloped surfaces or elevation changes, such as steps, slow down travel speed and break one's travel rhythm. Circulation across sloping surfaces should be minimized, or if that is unavoidable, travel should run at an angle to the contours, rather than perpendicular or straight up the slope. Finally, landforms may be arranged to force movement in a certain direction. For this purpose, one can use earth slopes or mounds to create barriers that force traffic around them and through the valley-like spaces.

SITE DESIGN CHARACTERISTICS

We have stated that site design is the conscious rearrangement of the environment for human use. The environment comprises a number of components, or characteristics, including space, scale, mass, proportion, etc., all of which contribute to that elusive and subjective element known as aesthetics. The aim of all site design, therefore, is to produce a functional solution that is perceived as a visual whole, that is, aesthetically. And aesthetics is the consequence of a harmonious blend of design characteristics.

Space

Space is defined as the three-dimensional expanse that surrounds one, and it is perceived through all our senses. While architectural space is circumscribed by roof, walls, and floor, outdoor space is defined by the variety of elements found in the open or urban landscape. This may include, for example, a group of university buildings around a quad, a narrow street lined with uniform structures, or an open park surrounded by trees. Some outdoor space is vast and limited only by sky, earth, and the distant horizon.

Compared to architectural space, site space is generally larger in extent, more irregular, less geometric, and invariably perceived as being

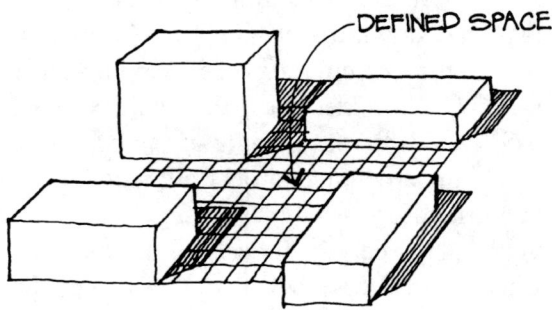

OUTDOOR SPACE DEFINED
BY BUILDING PLACEMENT

OUTDOOR SPACE DEFINED
BY ROWS OF TREES

wider than it is high. People relate to exterior space differently from the way they relate to interior space. In a vast, open plain, some will feel threatened or overwhelmed, while others will experience a sense of freedom or a need for action. Large areas often encourage mass action, such as on a football field or a ski slope. Any tall object set on a large, unobstructed surface becomes an important element on which attention is focused, such as a solitary tree in a field or the Washington Monument.

Space is further defined by light, color, texture, and the scale of its elements, although perceptions may be modified through spatial illusions. It is difficult, for example, to accurately estimate outdoor distances, and actual gradients may be considerably different from what they appear to

be. And outdoor light is not constant, since the sun varies with the hour, the season, and the weather, all of which affect the form, color, and texture of the landscape.

Enclosure

People are aware when they are inside a building, regardless of how open it may be, but outdoor enclosure is perceived differently. Outdoor spaces can be enclosed by widely spaced trees, rolling foothills, or a change in ground texture. In other words, the definition of outdoor space may be a visual suggestion, rather than a visual obstruction. Even a low railing or a line of bollards defines outdoor space as effectively as a wall. In general, the amount of enclosure necessary to create a definable outdoor space is just enough so that one's attention is focused on the space, rather than beyond it.

Urban spaces are enclosed by the building masses of a city; they are the voids formed by the absence of solids. An urban square surrounded by tall buildings is easy to visualize, but the linear arrangement of houses is no less an enclosure for the corridor-like space it creates along the street. Unlike indoor spaces, outdoor volumes may be infinite in scope, limited perhaps only by the horizon. Site designers, therefore, are not nearly as restricted as building designers. Because outdoor space is loosely defined, the site designer has greater freedom, as well as responsibility, to create a clear, comprehensible volume.

Scale

Scale is a system of relative measurement based on anthropometric dimensions. For example, a stair riser is about seven inches high, because that is the height that people raise their feet comfortably when climbing steps. We say, therefore, that a seven-inch riser is in scale. The 20-inch risers of the Parthenon's stylobate, on

CREATING URBAN SPACE

SCALE - PARTHENON STEPS

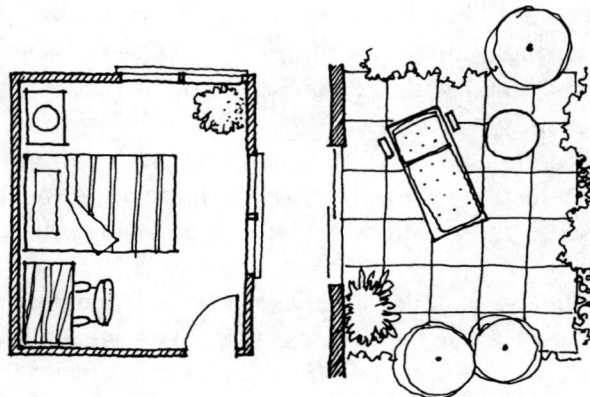

10 X 12 BEDROOM   10 X 12 PATIO

RELATIVE SCALE

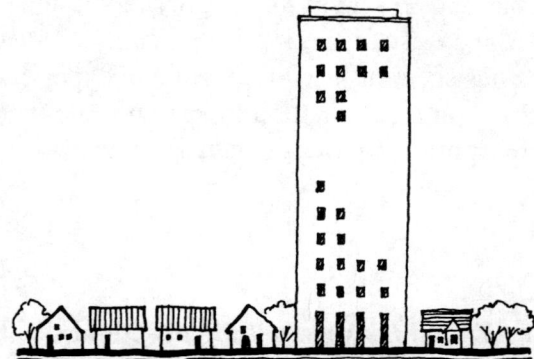

OUT OF CONTEXT / SCALE

the other hand, while in harmonious proportion to the structure, are clearly out of scale with human beings: people do not normally climb 20-inch high steps.

Interior scale and exterior scale are quite different from each other. In general, outdoor spaces must be considerably larger, relative to interior spaces, in order to feel comfortable. A three-foot wide hallway in a house, for example, feels adequate, whereas a three-foot wide sidewalk will feel narrow. Similarly, a 10 by 12 foot bedroom is comfortable, while the same size patio will be undersized. This variation in perception between inside and outside scale is caused by differences in the fields of vision, as well as the physical behavior that is appropriate in each circumstance. One can run, shout, or throw a ball outside, while these same activities inside are considered inappropriate.

Outside activities can be visually distinguished up to about 450 feet, outdoor spaces appear intimate if they are between 40 to 80 feet in size, and people who stand three to ten feet away are considered to be in direct relationship. It is clear, therefore, that site planning requires an entirely

different perception of scale than planning interior spaces.

Exterior scale is more than a matter of dimensions, however; it may also relate to speed, context, and custom. Walking speed, for example, which averages about 2-1/2 miles per hour, determines the size and scale of a city's elements. The willingness of people to walk only 10 to 15 minutes in performing routine tasks affects the arrangement of parking lots, shopping centers, commercial zones, and the size of neighborhoods. The relationship of scale to context means that neighboring spaces and buildings must be in scale with each other. A towering building in the midst of a community of single-family dwellings, for example, is out of context with the neighborhood, and consequently out of scale. All buildings in an area need not be the same mass or height, but when a structure changes one's perception of the local scale, the structure is likely to be disorienting.

Finally, scale is perceived in accordance with the customary way things have always been. If a coherent neighborhood has always consisted of 500 families who support a shopping center, recreational facilities, and an elementary school, then doubling the number of families would probably overwhelm the neighborhood and destroy the established scale. Similarly, a person who has spent his life on a farm will experience a dramatic disorientation in spatial scale if he were to move to a large metropolitan area.

Mass

The perception of mass is largely controlled by the way we see, as well as the prevailing light conditions. From a viewing distance equal to the building height, that is, a 45-degree angle from the eye to the roof line, one notices the details of the facade more than the entire building mass.

From twice the viewing distance to the same roof line, that is, a 1 to 2 relationship, one can perceive the entire building mass together with its details. At a 1 to 3 relationship, the building mass is observed in relation to surrounding objects, and at 1 to 4, or a viewing distance of four times the building height, one sees the mass as an edge that frames a distant view.

Light conditions affect one's perception of mass: in bright sunlight, individual elements stand out, while on cloudy days, the mass is perceived in its entirety. Dark objects seen against a light background recede, such as a tree clump viewed against the sky. However, light objects seen against a dark background, such as a highway billboard set against dark hills, tend to advance visually. Depth perception is also affected by light conditions: distances are more difficult to discern on dull, grey days.

Aesthetics

Aesthetics refers to what is beautiful, and beauty, as we all know, can be quite subjective. Most natural landscapes are beautiful, because the many factors of which they are composed have achieved an equilibrium. Beauty, therefore, must include a concept of order. A natural site may have existed for thousands of years, but it only exists in relation to people through some positive development which establishes a permanent connection between people and site. Even a natural wilderness area has no relation to people unless there is access to the area, or at least vantage points from which the area can be observed. A structure introduced on a site will exist visually and spatially in relation to that site and the surrounding landscape. Building and site become one indivisible experience. Unity, therefore, is a goal of the design process and an essential element of beauty, because it confirms our idea of what is right, proper, and fit for human use.

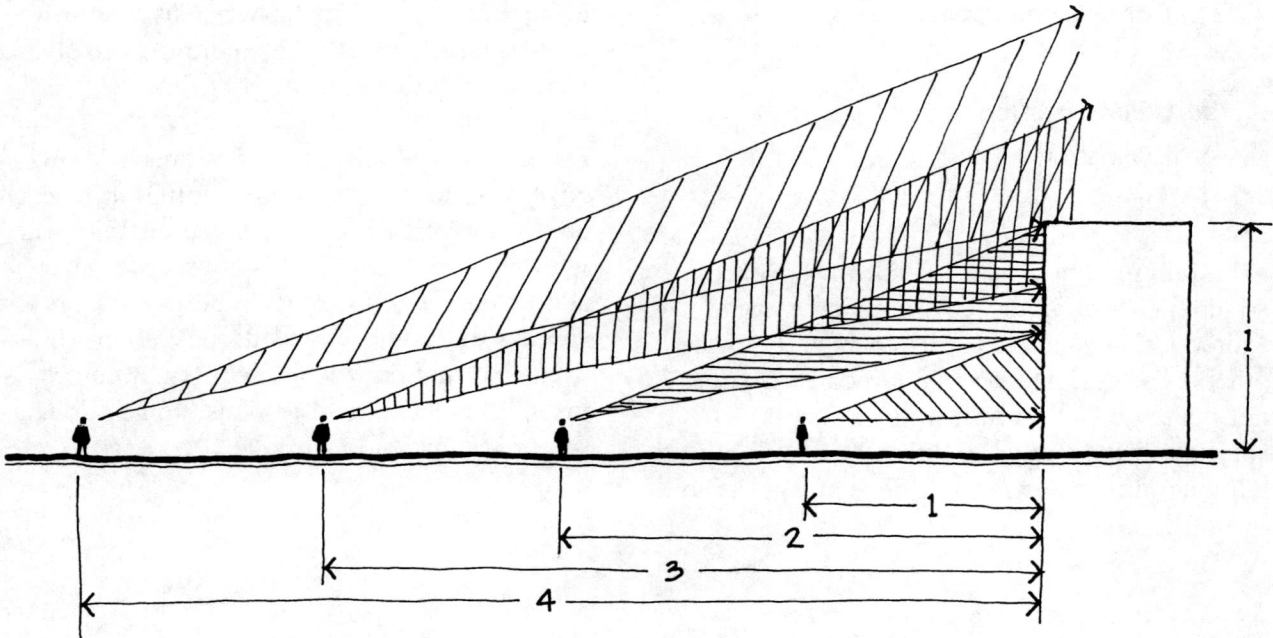

PERCEPTION OF MASS BASED ON VIEWING DISTANCE

SITE DESIGN PROCESS

The site design process is an exploration of possible solutions to a specific problem. This exploration involves a number of essential steps, generally performed in sequence, which ultimately leads to a solution of the project's objectives. In the usual case, a client intends to develop a piece of land for some purpose. The designer may be contacted by the client either before or after the site has been selected. Either way, the designer must become familiar with the client's goals, the intended land use, and the parcel of land itself. From that point on, the sequence of activities includes the following steps:

1. **Project Proposal**
    A. Scope of services
    B. Cost of services
    C. Time of performance

2. **Research and Analysis**
    A. Site inventory
    B. Data analysis
    C. Client objectives
    D. Program preparation

3. **Design Phase**
    A. Circulation pattern
    B. Functional pattern
    C. Form composition
    D. Diagrammatic plan
    E. Schematic plan
    F. Preliminary plan
    G. Master plan (design development)

4. **Construction Phase**
    A. Technical plan
    B. Grading plan
    C. Landscaping plan

D. Construction details

E. Contract documents

5. Post Construction

A. Evaluation

B. Maintenance

While these various steps occur in sequence, some may overlap or occur simultaneously. Moreover, no step occurs independently of the others. The design process outlined above does not guarantee a beautiful or even functional solution; it is merely a framework of activities that one must perform to achieve an answer to a specific puzzle. The answer may result in a masterpiece, a disaster, or more than likely, something in between.

Design success relies on a designer's knowledge, inspiration, experience, intuition, talent, ability, and creativity, and these qualities vary with the individual. In this course we have addressed one of the most important factors for successful design: knowledge. We hope, however, that this knowledge will lead to more inspired, creative, and responsible solutions to site problems.

LESSON 6 QUIZ

1. Throughout the history of urban communities, how did towns generally expand?

   A. Concentrically

   B. Rectilinearly

   C. Linearly

   D. Axially

2. Which word best describes urban planning during the Renaissance?

   A. Balanced

   B. Formal

   C. Radial

   D. Linear

3. Who was largely responsible for the dramatic transformation of Paris in the 19th century?

   A. Pierre L'Enfant

   B. Tony Garnier

   C. Viollet-le-Duc

   D. Baron Haussmann

4. In response to the deteriorating industrial environment, the details of the Garden City were developed by

   A. Daniel Burnham.

   B. Patrick Geddes.

   C. Ebenezer Howard.

   D. Clarence Stein.

5. Who conceived the "superblock," in an attempt to minimize conflicts between pedestrians and vehicles?

   I.   Daniel Burnham

   II. Ebenezer Howard

   III. Clarence Stein

   IV. Henry Wright

   A. I only

   B. I and II

   C. III only

   D. III and IV

6. The principal purpose of site analysis is to determine if a site

   A. requires modification.

   B. is suitable for a proposed use.

   C. is capable of being developed.

   D. has a unique character.

7. Outdoor space is perceived differently from indoor space, EXCEPT in the matter of

   A. scale.

   B. enclosure.

   C. proportion.

   D. mass.

8. The greatest degree of spatial enclosure is perceived with which of the following landforms?

   I.   Level

   II. Convex

   III. Concave

   IV. Ridges

   V.   Valleys

   A. V only

   B. I and IV

   C. II and III

   D. III and V

9. Development of a site always results in
   A. environmental modification.
   B. altered drainage patterns.
   C. increased pollution.
   D. damage to the ecological balance.

10. Which of the following site planning situations might be considered out of scale?
   A. A tall monument placed in an open area
   B. A small residential structure situated on top of a high summit
   C. An elevated freeway running through an area of small stores
   D. A high rise residential tower situated among commercial structures of equal height and mass

The following glossary defines a number of terms, many of which have appeared on past exams. While this list is by no means complete, it comprises much of the terminology with which candidates should be familiar. You are therefore encouraged to review these definitions as part of your preparation for the exam.

A

Absorption Field   See Disposal Field.

Access Right   Right of an owner to have ingress and egress to and from a property.

Accessory Building   A building or structure on the same lot as the main or principal building.

Active Solar System   A heating or cooling system that collects and moves solar heat with the assistance of mechanical power.

Aesthetics   The study or theory of beauty.

A Horizon   The layer of soil just below the surface.

Air Rights   The rights to the use or control of space above a property.

Albedo   Reflectivity measured as the relative permeability of a surface to radiant energy flowing in either direction.

Alignment   Horizontal or vertical deviation from the straight or level centerline of the road.

Altitude   The angle that the sun makes with the horizon.

Angle of Repose   The steepest angle with the horizontal at which loose earth will stand without sliding.

Aquifer   An underground permeable material through which water flows.

Area Drain   A device that collects water from the low point of a limited area and conducts it directly to underground pipes.

Autumnal Equinox   The beginning of autumn, about September 21, when night and day are of equal length.

Auxiliary Heat Source   A mechanical back-up heating system that is automatically activated when solar energy is insufficient to supply the required needs.

Azimuth   A horizontal angle measured clockwise from north or south.

B

Barrier-Free   Having no environmental barriers, thereby permitting free access and circulation by the handicapped.

Bearing   In surveying, a direction stated in degrees, minutes, and seconds as an angular deviation east or west from due north or south.

Bearing Capacity   The ability of a soil to support load.

Bench Mark   A relatively permanent point of known location and elevation.

Berm   A convex shaped bank of earth.

B Horizon   The soil layer just below the A horizon, consisting of weathered and decomposed rock material.

Boundary   The legal recorded property line between two parcels of land.

BTU   Abbreviation for British Thermal Unit: the amount of heat required to raise one pound of water one degree F.

Buildable Area   The net ground area of a lot that can be covered by a building after required setbacks and other zoning limitations have been accounted for.

Building Line   A defined limit within a property line beyond which a structure may not protrude.

C

Catch Basin   A drainage device used to collect water, with a deep pit to catch sediment.

Chemical Weathering   The change of rock composition by chemical processes. Also called Decomposition.

C Horizon   The layer of soil directly above bedrock, consisting of partially decomposed rock material.

Circulation   The flow or movement of people, goods, vehicles, etc., from place to place.

Clay   A fine-grained soil whose particles are smaller than 0.002 millimeters in diameter.

Climate   The generally prevailing weather conditions of a region throughout the year, averaged over a series of years.

Coefficient of Runoff   A fixed ratio of total rainfall that runs off a surface.

Collector Street   A street into which minor streets empty and which leads to a major arterial.

Combined Sewer   Sewer that carries both storm water and sanitary or industrial wastes.

Comfort Zone   Any combination of temperature and humidity in which the average person feels comfortable.

Compaction   The reduction of soil volume by pressure from grading machinery.

Condemnation   Taking private property for public use, with compensation to the owner, under the right of eminent domain.

Conditional Use   A use not strictly allowed in the zoning ordinance, but permitted if specified conditions are met and if approval has been granted by the local governing body.

Conduction   The transfer of heat by direct molecular action.

Conduit   Pipe or other channel, below or above ground, for conveying pipelines, cables, or other utilities.

Conforming Use   Lawful use of a building or lot that complies with the provisions of the applicable zoning ordinance.

Coniferous   Describing a cone-bearing tree or shrub. See Evergreen.

Contour   A line on a plan that connects all points of equal elevation.

Contour Interval   The vertical distance between adjacent contours.

Convection   The transfer of heat by the movement of a liquid or gas, such as air.

Covenant   A restriction of the deed that regulates land use, aesthetic qualities, etc., of an area.

Crown   The central area of a convex surface, such as a road.

Cul-De-Sac   A very short road with an outlet on one end and a turnaround on the other.

Culvert   A length of pipe under a road or other barrier used to convey water.

Curb   A raised margin running along the edge of a street pavement, usually of concrete.

Curvilinear Pattern   A circulation pattern comprised of curves, which closely follows the contours of the land.

Cut and Fill   In grading, earth that is removed (cut) or added (fill).

D

Deadman   An underground block of concrete or other material that is used as an anchor.

Deciduous   Describing trees that shed their leaves annually, as opposed to evergreen.

Decomposition   Chemical weathering of rock.

Dedication   Appropriation of private property for public use, together with acceptance for such use by a public agency.

Deed   A written instrument that is used to transfer real property from one party to another.

Degree Days   The number of degrees that the mean temperature for any day at a particular location is below 65°F.

Density   A measure of the number of people, families, etc., that occupy a specified area.

Discharge   Flow from a culvert, sewer, channel, etc.

Disintegration   Mechanical weathering of rock.

Disposal Field   A system of trenches with gravel and loose pipes through which septic tank effluent may seep into the surrounding soil. Also called Drainage Field or Absorption Field.

District   (1) Any section of a city in which the zoning regulations are uniform. (2) A section of the environment that has an identifying character.

Drainage   (1) The capacity of a soil to receive and transmit water. (2) The system by which excess water is collected, conducted, and dispersed.

Drainage Field   See Disposal Field.

Dwelling Unit   An independent living area that includes its own private cooking and bathing facilities.

E

Earthwork   See Grading.

Easement   A limited right, whether temporary or permanent, to use the property of another in a certain way. This may include the right of access to water, light and air, right-of-way, etc.

Ecology   The study of the pattern of relations between organisms and their environment.

Edge   The boundary between two districts. (See District, definition 2.)

Effective Temperature   The sensation produced by the combined effects of temperature, relative humidity, and air movement.

Effluent   Partially treated liquid sewage flowing from any part of a disposal system to a place of final disposition.

Elevation   The vertical distance above sea level or other known point of reference.

Eminent Domain   The right of a government, under the police power concept, to take private property for public use.

Encroachment   Part of a building or an obstruction that extends into the property of another.

Environment   The natural and man-made things, conditions, and influences surrounding a person, community, or place.

Environmental Impact Statement   A statement, often required by a governmental body, that assesses the environmental impact of a proposed development.

Equinox   See Autumnal Equinox and Vernal Equinox.

Erosion   The process by which the surface of the earth is worn away by the action of natural elements, such as water and wind. Also known as Weathering.

Evergreen   Having green leaves throughout the year, as opposed to deciduous.

Excavation   The digging or removal of earth.

Expansive Soil   Clay that swells when wet and shrinks when dried.

F

Finish Grade   The elevation of the ground surface after completion of all work.

Flat Plate Collector   A device used to collect solar energy.

Flood Plain   The land surrounding a flowing stream over which water spreads when a flood occurs.

Floor Area Ratio (FAR)   The ratio of the floor area of a building to the area of the lot.

Flow Line   The path in which water flows down.

Fossil Fuel   A natural fuel, such as oil, coal, or natural gas, formed from the remains of pre-historic plants or animals.

Frontage   The length of a lot line along a street or other public way.

Frost Line   The deepest penetration of frost below grade.

G

Geology   The science that deals with the physical history of the earth.

Grade   The elevation of any point. See also Gradient and Grading.

Gradient   The rate of slope between two points on a surface, determined by dividing their vertical difference in elevation by their horizontal distance apart.

Grading   The modification of earth to create landforms.

Gravel   A coarse-grained soil whose particles are larger than 2.0 millimeters in diameter.

Greenbelt   A belt-like area around a city, reserved by ordinance for park land, farms, open space, etc.

Greenhouse Effect   The direct gain of solar heat, generally through south-facing glass walls and roofs.

Grid Pattern   A pattern of circulation named for its shape.

Ground Water Level   The plane below which the soil is saturated with water. Also called Ground Water Table or Water Table.

H

Hachure   A shading technique used to depict ground form.

Handicapped   Describing individuals with physical impairments that result in functional limitations.

Handicapped Parking   A space designated for physically handicapped persons, consisting of a typical space with adjacent access aisle no less than five feet wide.

Humidity   The amount or degree of moisture in the air.

Humus   The dark organic material produced by the decomposition of living matter and essential to a soil's fertility.

Hydrologic Cycle   See Water Cycle.

I

Igneous Rock   Rock formed when molten rock material cools and solidifies on or beneath the earth's surface.

Impervious Soil   Tight cohesive soil, such as clay, which does not allow the ready passage of water.

Indigenous   Native.

Infiltration   The process by which water soaks into the ground. Also called Percolation.

Insolation   The amount of solar radiation on a given plane.

Interchange   The junction of a freeway with entering or exiting traffic.

Interpolation   Determining an unknown value between known values.

Intersection   The point at which two streets come together or cross.

Invert Elevation   The elevation of the bottom (flow line) of a pipe.

L

Land Coverage   The ratio of the area covered by buildings to the total lot area, expressed as a percentage.

Landform   The shape of the earth's surface.

Landmark   A prominent visual feature that acts as a point of reference.

Landscaping   The conscious rearrangement of natural outdoor elements for function and pleasure.

Latitude   The number of degrees north or south of the equator of a particular point on the earth's surface.

Legal Description   Designation of boundaries of real estate in accordance with one of the systems prescribed by law.

Linear Pattern   A pattern of land use which develops along a line, such as a highway or river.

Lithification   The process by which deposited sediments are converted to firm rock.

Loop Street   A minor street which comes off a major street, runs for a short distance, and then returns to the major street.

Lot Line   The boundary line of a lot.

Lot Area   Total horizontal area within the lot lines of a parcel of land.

M

Macroclimate   The general climate of a region.

Manhole   An access hole in a drainage system to allow inspection, cleaning, and repair.

Mechanical Weathering   The breaking of rock into smaller fragments by physical forces. Also called Disintegration.

Metamorphic Rock   Rock formed from igneous or sedimentary rock as a result of heat, pressure, and chemical action.

Metamorphism   The process by which igneous or sedimentary rock is converted to metamorphic rock.

Metes and Bounds   A formal description of the boundary lines of a parcel of real property in terms of the length and direction of those lines.

Microclimate   The climatic characteristics unique to a small area, caused by local features.

Multiple Dwelling   A building containing three or more dwelling units.

N

Neighborhood   A community of people living in a general vicinity. The area can generally support an elementary school.

Network   A system of circulation channels which covers a large area.

Node   A center of activity, such as a square or plaza.

Non-Conforming Use   A particular use of land or a structure that is in violation of the applicable zoning ordinance. Generally, if the use was established prior to the ordinance that it contravenes, it may continue to exist.

O

Off-Street Parking   Space provided for vehicular parking outside the dedicated street right-of-way.

Open Drainage   The removal of unwanted water by means of surface devices.

Orientation   A position with respect to the points of the compass.

P

Pad   An approximately level building area.

Party Wall   A wall built on the dividing line between two adjoining parcels, in which each owner has an equal share of ownership.

Passive Solar System   A heating or cooling system that collects and moves solar heat without using mechanical power.

Path   A circulation route along which people move.

Percolation   See Infiltration.

Planting Strip   A landscaped strip of ground dividing a pedestrian walk from a street.

Police Power The legal power of a government to authorize actions that are in the best interest of the general public.

Precipitation Water that falls on the land as rain or snow.

Principal Building A building that houses the main use or activity occurring on a lot or parcel of ground.

Property Line A legal boundary of a land parcel.

PUD A planned unit development, similar to a cluster development but larger in scale including, in addition to housing, commercial and industrial developments.

R

Radial Pattern A circulation pattern in which channels spread out from a central point.

Radiation The process in which heat or other energy is emitted by a body, transmitted through space, and absorbed by another body.

Rational Method A method for computing approximate storm water runoff.

Relative Humidity The ratio of the actual amount of moisture in the air to the maximum amount of moisture the air could hold at a given temperature.

Restrictions Limitations on the use of property defined by covenant in deeds, by private agreement, or by public legislative action.

Retaining Wall A wall that resists the lateral pressure of the earth behind it.

Ridge A linear convex landform, represented by contours pointing downhill.

Right-of-Way A strip of land granted by deed or easement for a circulation path.

Ring Pattern A land use pattern that is developed in a circular or doughnut form.

Runoff The surface flow of water from an area.

S

Sand A coarse-grained soil whose particles are 2.0 to 0.05 millimeters in diameter.

Sanitary Sewer An underground pipe or drain used to carry off waste matter.

Sedimentary Rock Rock formed by the deposition of transported sediments.

Septic System A sewage treatment system consisting of a tank and filtering system.

Setback The minimum legal distance between a property line and a structure.

Sewer An underground pipe or drain used to carry off excess water and/or waste matter. See Sanitary Sewer and Storm Drain.

Sheeting A thin layer of water moving across a surface. Also called Sheet Flow.

Silt A fine-grained soil whose particles are 0.05 to 0.002 millimeters in diameter.

Site Planning The art or science of creating or arranging the external physical environment.

Slope The inclination of a surface expressed as a percentage or proportion.

Sludge Accumulated solids that settle out of the sewage, forming a semi-liquid mass on the bottom of a septic tank.

Soil A natural material, formed of decomposed and disintegrated parent rock, that supports plant life.

Soil Boring Log A graphic representation of the soils encountered in a test boring.

Solar Zoning An ordinance controlling the mass and shape of buildings, to permit the penetration of sunlight between buildings.

Solstice See Summer Solstice and Winter Solstice.

Split Lot A lot that comprises more than one zone.

Spot Elevation The exact elevation at a key point on the ground or on a structure.

Spot Zoning   Zoning that differs from the pattern of the surrounding area.

Storm Drain   An underground pipe or drain used to carry off surface rain water.

Story   The vertical portion of a building included between the surface of any floor and the surface of the floor next above.

Subsidence   The sinking of land.

Summer Solstice   The beginning of summer, about June 21, which is also the longest day of the year.

Summit   The highest point of a land mass, represented by concentric contours.

Sun Chart   A map of the sky showing the path of the sun, from sunrise to sunset, on the 21st day of each month.

Superblock   A large area of land in which all through traffic is eliminated, but which may include cul-de-sacs.

Surcharge   Vertical load or a sloping ground surface behind a retaining wall, which causes increased earth pressure.

Swale   A graded flow path used in open drainage systems.

Switchback Road   A road that doubles back on itself with a hairpin curve.

T

Test Boring   A hole drilled into the ground, from which samples of undisturbed subsurface soils are obtained for laboratory inspection and testing.

Test Pit   An excavation made to expose the subsurface soils for in-place examination.

Topography   The configuration of the earth's surface.

Topsoil   The upper six to eight inches of soil, which contains humus.

Transpiration   The process by which water vapor escapes into the atmosphere from plants.

Trench Drain   A linear drainage device used to collect and conduct water.

V

Valley   A linear concave landform represented by contours pointing uphill.

Variance   The special permission granted to the owner of a parcel of land waiving certain specific restrictions when the enforcement of these would impose an unusual or unreasonable hardship on the owner.

Vernal Equinox   The beginning of spring, approximately March 21, when night and day are of equal length.

W

Water Cycle   The general pattern of movement of the water on, under, and above the earth. Also called Hydrologic Cycle.

Water Table   See Ground Water Level.

Way   Street, alley, or other thoroughfare or easement permanently established for passage of persons or vehicles.

Weathering   See Erosion.

Windbreak   A structure or plant which, because of its form and location, reduces wind velocities.

Wind Shadow   The area near the bottom of the leeward side of a hill, where the wind velocity decreases to almost zero.

Winter Solstice   The beginning of winter, about December 21, which is also the shortest day of the year.

Y

Yard   Open, unoccupied space on all sides of a building, based on the required setbacks.

Z

Zone   Area established by a governing body for specific use, such as residential, commercial, or industrial use.

Zone of Aeration   The zone below the ground in which the spaces between soil grains contain both water and air.

Zone of Saturation   The zone below the ground in which all of the spaces between soil grains are filled completely with water.

Zoning   The legal means whereby land use is regulated and controlled for the general welfare.

Zoning Envelope   The volume within which a building may legally be placed.

Zoning Ordinance   A law or regulation by which a government exercises its police power in regulating and controlling the character and use of property.

BIBLIOGRAPHY

The following list of books is provided for candidates who may wish to do further research or study in Site Design. Most of the books listed below are available in college or technical bookstores, and all would make welcome additions to any architectural bookshelf. In addition to the course material and the volumes listed below, we advise candidates to review regularly the many professional journals, which are available at most architectural offices.

ANSI A117.1 Handicapped Standards
American National Standards Institute
New York, New York

Architectural Graphic Standards
Ramsey and Sleeper
John Wiley & Sons, Inc.
New York, New York

Basic Elements of Landscape Architectural Design
Booth, Norman K.
Elsevier Science Publishing Co.
New York, New York

Design with Climate
Olgyay, Victor
Princeton University Press
Princeton, New Jersey

Environmental Analysis
Marsh, William M.
McGraw-Hill Book Co.
New York, New York

Landscape Architecture
Simonds, John O.
McGraw-Hill Book Co.
New York, New York

Landscape Planning for Energy Conservation
Environmental Design Press
Reston, Virginia

Principles & Practices of Grading, Drainage, and Road Alignment
Untermann, Richard K.
Reston Publishing Co.
Reston, Virginia

Site Planning
Lynch, Kevin
M.I.T. Press
Cambridge, Massachusetts

Site Planning Standards
De Chiara, Joseph
McGraw-Hill Book Co.
New York, New York

Solar Dwelling Design Concepts
AIA Research Corporation
Washington, DC

The Passive Solar Energy Book
Mazria, Edward
Rodale Press
Emmaus, Pennsylvania

The Urban Pattern
Gallion, Arthur B.
Van Nostrand Reinhold Co.
New York, New York

Time Saver Standards for Site Planning
DeChiara/Koppelman
Van Nostrand Reinhold Co.
New York, New York

Urban Design
Spreiregen, Paul D.
McGraw-Hill Book Co.
New York, New York

QUIZ ANSWERS

Lesson 1

1. **D** In the winter, the sun is low and to the south. Therefore, the south side of a building receives maximum solar radiation during that season. See page 6.

2. **B** Deciduous trees (I) shade the summer sun, but allow the winter sun to penetrate. Similarly, a south overhang (IV) is effective in blocking the high summer sun, while allowing the low winter sun to enter.

3. **B** See page 6.

4. **A** See page 17.

5. **C** South-facing slopes receive more solar radiation than level or north-facing slopes. Therefore, they are comparable to level sites in more southerly latitudes.

6. **C** Windward slopes (IV) tend to be cool, humid, and vegetated (see page 4). North-facing slopes (II) receive little solar radiation and therefore also tend to be cool, humid, and often vegetated.

7. **D** Latitude, proximity to water, and elevation all affect the macroclimate. Windbreaks affect only the microclimate.

8. **C** Preferred sites for residential development include those near a body of water, because of its moderating influence, and those sheltered from cold northerly winds. Sites at the bottom of a hill tend to be windy, making such sites less desirable. Summer breezes generally come from the southwest; therefore, sites exposed to such breezes are preferable to those sheltered from them.

9. **B** Referring to the sun chart on page 7, one can determine all of the factors in this question except the number of sunny days during the year.

10. **B** See page 10.

Lesson 2

1. **B** See the table on page 30.

2. **A** See page 28.

3. **C** The description in choice A applies to topsoil, not C horizon. B is incorrect because erosion has not taken place at a constant rate, but has been accelerated by man-made developments. And humid climates tend to have deep, not shallow, soil.

4. **C** See page 43.

5. **D** Building foundations may extend into the zone of saturation, basement walls must be designed for hydrostatic pressure if the ground water table is high, regardless of waterproofing or gravel backfill, and the ground water table is usually a sloping surface. See page 45.

6. **D** See page 45.

7. **A** The most effective barriers against unwanted sound are structures, walls, or earth, rather than trees.

8. **C** See pages 48 and 50.

9. **B** See page 43.

10. **D** Expansive soils swell when wet and shrink when dried. Footings in such soils should be extended below the depth of seasonal moisture change, so that the moisture content of the subsoil remains fairly constant. Either spaced foundation piers or footings may be used in expansive soils.

Lesson 3

1. **B** Contours spaced at increasing intervals going uphill, as in this case, indicate a convex slope.

2. **A** Contours pointing uphill, as at B, indicate a valley or swale.

3. **D** Existing contour 40 is roughly halfway between proposed contours 40 and 42. In other words, the elevation will be raised from 40 to about 41, requiring about one foot of fill.

4. **C** The drainage at D is from elevation 36 towards contour 34, which is northwest.

5. **A** At E, the elevation rises from 30 to 36 in a horizontal distance of 40 feet. The gradient is $V \div H = 6 \div 40 = 0.15$, which is 15 percent.

6. **A** Cut earth is generally more stable than filled earth (I), and therefore, cut slopes are generally permitted to be steeper than fill slopes (III).

7. **C** See page 62.

8. **C** Spot elevations are exact elevations of key points on (or under) the ground, such as inverts, tops of curbs, and building corners.

9. **D** See page 76.

10. **B** Storm drainage systems are designed to achieve all the factors listed, except the improvement of soil texture. See page 66.

Lesson 4

1. **A** See page 103.

2. **B** All of the statements are correct except choice B. For proper drainage, the ground surface, including planted areas, must be sloped considering the volume of water expected, the nature of the ground surface, and the potential damage that could be caused by flooding.

3. **C** Paved parking areas should be sloped to drain, but not more than 5 percent in any direction.

4. **A** Crushed rock is considered to be a soft surface suitable for areas with light pedestrian traffic, such as a nature path.

5. **D** Statements II and III are correct: there is no consideration more important than public safety (II), and road alignments that parallel existing contours minimize the amount of grading required (III). Access locations close to major intersections are undesirable, because driveways interfere with traffic in the intersection (I is incorrect), and urban sites are usually accessible from streets along one property line, which limits the number of choices for locating points of access (IV is incorrect).

6. **B** Water supply systems are under pressure and therefore do not need to be sloped (A is incorrect). The distribution system may be connected to the user by either a public or private water company (C is incorrect), and most systems distribute water for domestic, industrial, and fire fighting use from a single network (D is incorrect).

7. **C** See page 102.

8. **B** See page 98. Non-ambulatory refers to a person's inability to walk. Pedestrians prefer to walk on the most direct paths, not necessarily those laid out in a rigid geometric pattern. And three square feet per person is inadequate space for an outdoor assembly area, because it would permit no movement. See page 97.

9. **A** An interchange is often located where an expressway overpass crosses an arterial highway. Other uses should preferably be located at some distance from the interchange.

10. **D** Storm drains are usually provided by a public agency, they are often about five feet below grade, and they are more critical in urban than rural areas because of the greater amount of runoff.

Lesson 5

1. **D** See page 115.

2. **C** Density is defined as the number of people or families that occupy a given area. One way to control density is to limit the number of dwelling units that may be built on a given parcel of land.

3. **D** Zoning ordinances may limit the number of stories in a building and/or the building's height, as well as prescribe setbacks, in order to maintain a certain amount of sky exposure.

4. **A** See page 118.

5. **B** See page 125.

6. **B** See page 120.

7. **A** Cluster zoning allows higher density of housing in exchange for common open space. The overall density, however, is the same as for a conventional development.

8. **A** See page 124.

9. **D** Incentive or bonus zoning may benefit the city, the owner, and the tenants. See page 122.

10. **D** Foundation costs are affected by various factors, including soil conditions, topography, depth of water table, and the size and design of the building. Vegetation, however, has little or no effect on foundation costs.

Lesson 6

1. **A** See page 136.

2. **B** Urban planning during the Renaissance was intellectual, formal, and monumental, and spacious plazas, boulevards, and gardens were introduced in order to provide open space within the city.

3. **D** See page 138.

4. **C** Ebenezer Howard developed the details of the Garden City, consisting of a central core of public buildings surrounded by commercial shops and dwellings, with industrial facilities on the outskirts. All of this was to be surrounded by a permanent green belt.

5. **D** Inspired by the Garden City concept, Clarence Stein and Henry Wright developed the "superblock," which would minimize vehicle-pedestrian conflict.

6. **B** See page 141.

7. **C** Outdoor space is perceived differently from indoor space in scale, enclosure, and mass. Proportion, however, involves harmony between component elements and is perceived similarly indoors and outdoors.

8. **D** Concave landforms and valleys have a high degree of spatial enclosure, because they are defined by surrounding or adjacent slopes.

9. **A** The specific environmental changes in choices B, C, and D may or may not occur in any given site development. However, all site development results in some environmental modification.

10. **C** An elevated freeway in an area of small stores is out of context with the neighborhood and consequently out of scale.

The examination on the following pages should be taken when you have completed your study of all the lessons in this course. The questions have been designed to be similar to actual exam questions. Many questions are intentionally difficult in order to reflect the pattern of questions you may expect to encounter on the actual examination.

You will also notice that the subject matter for several questions has not been covered in the course material. This situation is inevitable and, thus, should provide you with practice in making an educated guess. Other questions may appear ambiguous, trivial, or simply unfair. This too, unfortunately, reflects the actual experience of the exam and should prepare you for the worst you may encounter.

Answers and complete explanations will be found on the pages following the examination, to permit self-grading. **Do not look at these answers until you have completed the entire exam**. Once the examination is completed and graded, your weaknesses will be revealed, and you are urged to do further study in those areas.

Please observe the following directions:

1. The examination is closed book; please do not use any reference material.

2. Allow about 1-1/2 hours to answer all questions. Time is definitely a factor to be seriously considered.

3. Read all questions *carefully* and mark the appropriate answer on the answer sheet provided.

4. Answer all questions, even if you must guess. Do not leave any questions unanswered.

5. If time allows, review your answers, but do not arbitrarily change any answer.

6. Turn to the answers only after you have completed the entire examination.

GOOD LUCK!

EXAMINATION ANSWER SHEET

Directions: Read each question and its lettered answers. When you have decided which answer is correct, blacken the corresponding space on this sheet. After completing the exam, you may grade yourself; complete answers and explanations will be found on the pages following the examination.

1 Ⓐ Ⓑ Ⓒ Ⓓ
2 Ⓐ Ⓑ Ⓒ Ⓓ
3 Ⓐ Ⓑ Ⓒ Ⓓ
4 Ⓐ Ⓑ Ⓒ Ⓓ
5 Ⓐ Ⓑ Ⓒ Ⓓ
6 Ⓐ Ⓑ Ⓒ Ⓓ
7 Ⓐ Ⓑ Ⓒ Ⓓ
8 Ⓐ Ⓑ Ⓒ Ⓓ
9 Ⓐ Ⓑ Ⓒ Ⓓ
10 Ⓐ Ⓑ Ⓒ Ⓓ
11 Ⓐ Ⓑ Ⓒ Ⓓ
12 Ⓐ Ⓑ Ⓒ Ⓓ
13 Ⓐ Ⓑ Ⓒ Ⓓ
14 Ⓐ Ⓑ Ⓒ Ⓓ
15 Ⓐ Ⓑ Ⓒ Ⓓ
16 Ⓐ Ⓑ Ⓒ Ⓓ
17 Ⓐ Ⓑ Ⓒ Ⓓ
18 Ⓐ Ⓑ Ⓒ Ⓓ
19 Ⓐ Ⓑ Ⓒ Ⓓ
20 Ⓐ Ⓑ Ⓒ Ⓓ
21 Ⓐ Ⓑ Ⓒ Ⓓ
22 Ⓐ Ⓑ Ⓒ Ⓓ
23 Ⓐ Ⓑ Ⓒ Ⓓ
24 Ⓐ Ⓑ Ⓒ Ⓓ
25 Ⓐ Ⓑ Ⓒ Ⓓ
26 Ⓐ Ⓑ Ⓒ Ⓓ
27 Ⓐ Ⓑ Ⓒ Ⓓ
28 Ⓐ Ⓑ Ⓒ Ⓓ
29 Ⓐ Ⓑ Ⓒ Ⓓ
30 Ⓐ Ⓑ Ⓒ Ⓓ

31 Ⓐ Ⓑ Ⓒ Ⓓ
32 Ⓐ Ⓑ Ⓒ Ⓓ
33 Ⓐ Ⓑ Ⓒ Ⓓ
34 Ⓐ Ⓑ Ⓒ Ⓓ
35 Ⓐ Ⓑ Ⓒ Ⓓ
36 Ⓐ Ⓑ Ⓒ Ⓓ
37 Ⓐ Ⓑ Ⓒ Ⓓ
38 Ⓐ Ⓑ Ⓒ Ⓓ
39 Ⓐ Ⓑ Ⓒ Ⓓ
40 Ⓐ Ⓑ Ⓒ Ⓓ
41 Ⓐ Ⓑ Ⓒ Ⓓ
42 Ⓐ Ⓑ Ⓒ Ⓓ
43 Ⓐ Ⓑ Ⓒ Ⓓ
44 Ⓐ Ⓑ Ⓒ Ⓓ
45 Ⓐ Ⓑ Ⓒ Ⓓ
46 Ⓐ Ⓑ Ⓒ Ⓓ
47 Ⓐ Ⓑ Ⓒ Ⓓ
48 Ⓐ Ⓑ Ⓒ Ⓓ
49 Ⓐ Ⓑ Ⓒ Ⓓ
50 Ⓐ Ⓑ Ⓒ Ⓓ
51 Ⓐ Ⓑ Ⓒ Ⓓ
52 Ⓐ Ⓑ Ⓒ Ⓓ
53 Ⓐ Ⓑ Ⓒ Ⓓ
54 Ⓐ Ⓑ Ⓒ Ⓓ
55 Ⓐ Ⓑ Ⓒ Ⓓ
56 Ⓐ Ⓑ Ⓒ Ⓓ
57 Ⓐ Ⓑ Ⓒ Ⓓ
58 Ⓐ Ⓑ Ⓒ Ⓓ
59 Ⓐ Ⓑ Ⓒ Ⓓ
60 Ⓐ Ⓑ Ⓒ Ⓓ
61 Ⓐ Ⓑ Ⓒ Ⓓ

FINAL EXAMINATION

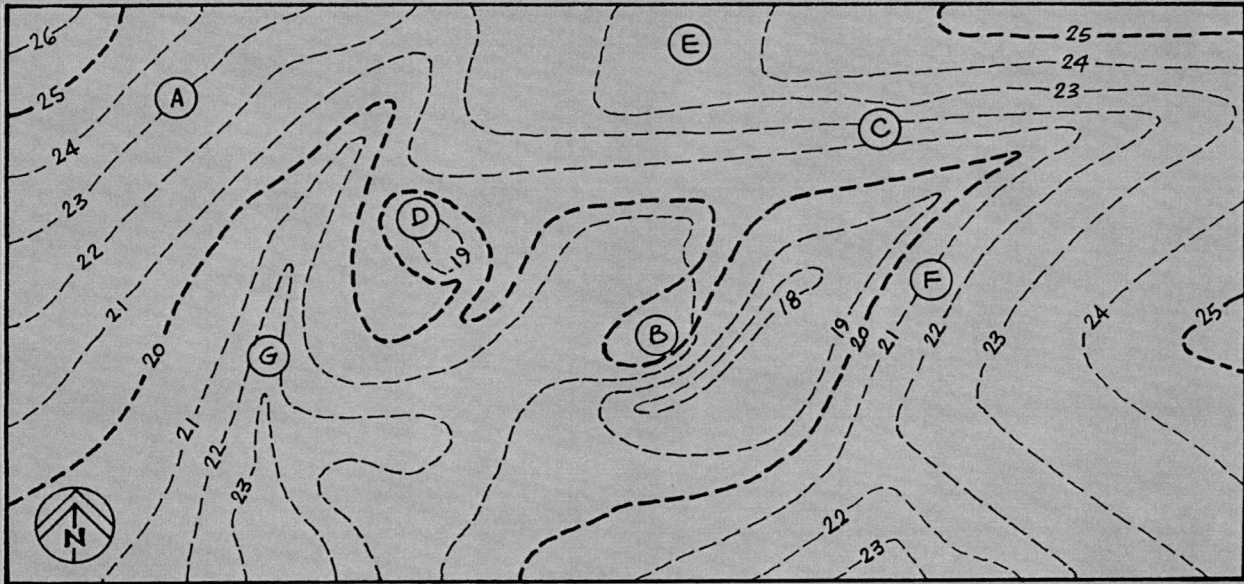

Questions 1 to 7 refer to the topographic map above.

1. The slope at point A is
   A. concave.   C. uniform.
   B. convex.   D. steep.

2. What is the configuration at point B?
   A. Depression
   B. Overhanging cliff
   C. Summit
   D. Valley

3. In what direction is the drainage at point C?
   A. Northerly   C. Southerly
   B. Easterly   D. Westerly

4. The landform at point D is a
   A. depression.   C. valley.
   B. summit.   D. ridge.

5. How would you describe the topography at point E, compared to all other points on the map?
   A. Lower   C. Steeper
   B. Higher   D. Flatter

6. The slope at point F can be described as
   A. concave.   C. uniform.
   B. convex.   D. east-facing.

7. What is the configuration at point G?
   A. Ridge   C. Swale
   B. Valley   D. Summit

8. On a road with a gradient of 5 percent, what will the elevation be 150 feet downhill from a point at elevation 132.5?
   A. 17.5   C. 125.9
   B. 125.0   D. 140.0

9. In comparison to a two-way street, a one-way street
   A. increases the rate of traffic flow.
   B. increases the length of utility connections.
   C. results in fewer intersections.
   D. is safer for pedestrians.

10. Where the terrain is steep and irregular the installation of utilities for the development of a site will be
    A. considerably more difficult than for a dead flat site.
    B. more costly than for a level site because it will require longer runs and manholes at more frequent intervals.
    C. less costly because the excavation for conduits will be less than for a level site.
    D. more costly because of the many changes in the horizontal direction.

11. One of the principal reasons that Medieval fortified towns were built in a circular pattern was that
    A. most were built on irregular terrain, and the walls followed the natural contours.
    B. Roman technology was lost during the Dark Ages, and right angles were difficult to construct.
    C. the heavy, protective surrounding walls enclosed the most area for the least effort and expense.
    D. most existing towns were circular in shape, thereby establishing a pattern for the enclosing walls.

12. The use of thick walls, high ceilings, wide overhangs, and limited fenestration is typical in which climatic zone in the United States?
    A. Hot-humid   C. Temperate
    B. Hot-arid   D. Cool

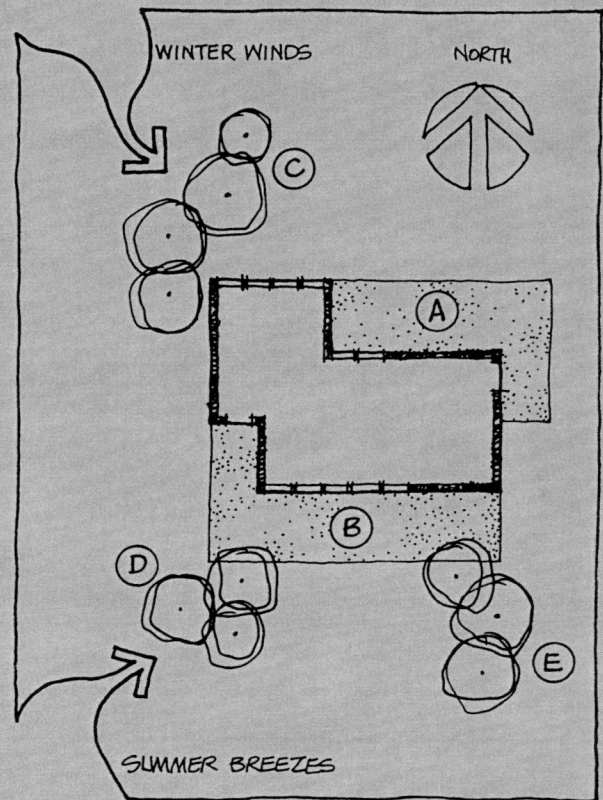

Questions 13 to 15 refer to the residential site shown above, located at 40° North Latitude.

13. Select the correct statement. Terrace A
    A. is ideally located for winter use.
    B. is ideally located for summer use.
    C. receives morning sun mainly in the winter.
    D. receives afternoon sun all year round.

14. What do trees C do?

I.   Provide protection from the winter wind if they are deciduous.

II. Provide protection from the winter wind if they are evergreen.

III. Provide protection from the afternoon summer sun.

IV. Provide protection from the afternoon winter sun.

A. I and III   **C.** I and IV

B. II and III   **D.** II and IV

15. Select the correct statements.

I.   Trees C should be evergreen.

II. Trees D should be evergreen.

III. Trees E should be evergreen.

A. I and II   **C.** I only

B. II and III   **D.** I, II, and III

16. The cost of grading can be reduced by all of the following EXCEPT

A. balancing the amounts of cut and fill.

B. planning grading operations to avoid winter work.

C. reducing the amount of land covered by buildings.

D. saving all mature trees and shrubs.

17. Open storm drainage systems are preferred to closed systems because

A. they are less expensive.

B. they have a greater capacity.

C. their gradients are more flexible.

D. they eliminate standing water.

18. Solid contour lines do which of the following?

I.   Connect points of equal elevation.

II. Represent proposed landform modifications.

III. Indicate natural topographic configurations.

IV. Never close on themselves.

V.   Never split in two.

A. I and V   **C.** I, II, and V

B. III and IV   **D.** I, II, IV, and V

19. Select the correct statement about solar energy.

A. Flat plate collectors absorb only direct radiation.

B. Flat plate collectors should ideally be oriented due south, but a variation up to 45° either side of due south will not appreciably reduce the collector's efficiency.

C. Water is the most common medium for heat transfer and storage in solar systems.

D. Supplying 100 percent of the hot water requirement by solar heating is usually economically feasible.

20. On a slope, where is the vegetation likely to be thickest?

A. Near the toe of the slope

B. At midslope

C. At the top of the slope

D. Where each soil layer plane intersects the slope

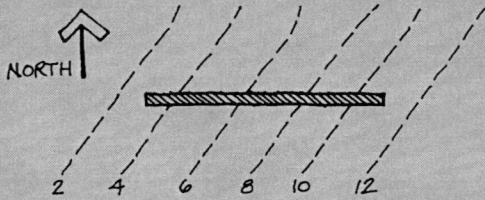

21. The sketch above shows the plan of a proposed retaining wall on naturally sloping ground. If the area on the north side of the wall were graded to a level elevation of 4, what would be the elevation necessary at the highest point along the wall's top?

A. 4 feet	**C.** 9 feet
B. 7 feet	**D.** 11 feet

Questions 22 through 27 are based on the site survey shown in the next column and the development standards and building program shown below.

Development Standards

Setbacks: 25' at north and south property lines
    20' at east and west property lines for all floors above 2nd

Height Limit: 11 stories above grade

Maximum Lot Coverage: 60%

Floor Area Ratio (FAR): 3.0 (excluding garages)

Off-street Parking: 400 s.f. for every 1000 s.f. of building area

Building Program

An office building having the maximum allowable floor area.

Floor to floor height is 12'-0".

No projections above the roof.

Off-street parking below grade.

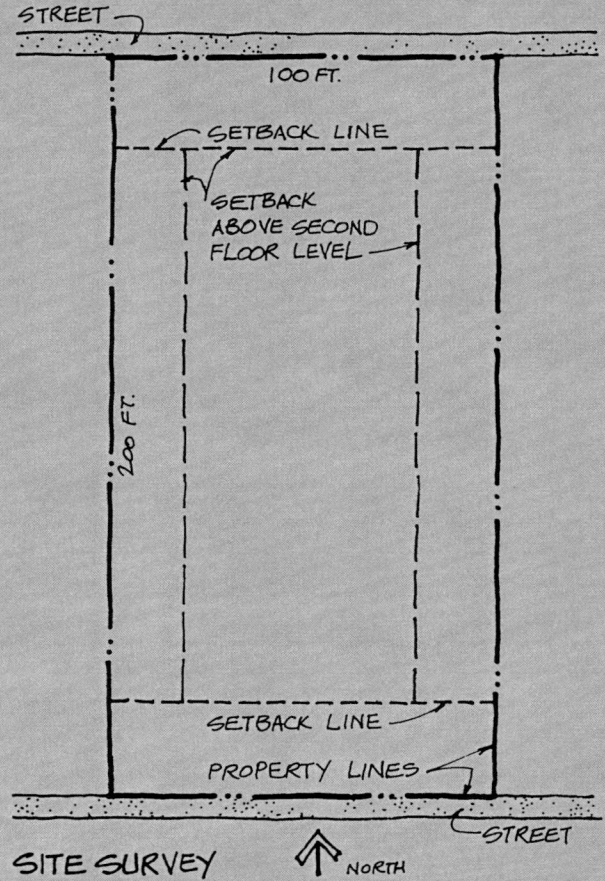

SITE SURVEY   NORTH

22. What is the maximum ground floor area of the office building?

    A. 5,400 square feet

    B. 9,000 square feet

    C. 12,000 square feet

    D. 15,000 square feet

23. The gross building area, exclusive of garages, may not exceed

   A. 60,000 square feet.

   B. 99,000 square feet.

   C. 111,000 square feet.

   D. 132,000 square feet.

24. If the office building has the maximum ground floor area and the maximum gross building area, how many levels of underground parking are required? Assume that the parking structure is located directly below, and has the same configuration as, the ground floor of the building.

   A. One   **C.** Three

   B. Two   **D.** Four

25. Considering the allowable lot coverage and the required setbacks, what is the minimum number of floors that will accommodate the maximum allowable area of the building?

   A. Four   **C.** Six

   B. Five   **D.** Seven

26. If the first two stories utilize the entire area within the allowable lot coverage and the building is 11 stories in height, the tower above the second story would be nine stories high. What would be the maximum area of each floor in the tower?

   A. 3,333 sq. ft.   **C.** 4,667 sq. ft.

   B. 4,000 sq. ft.   **D.** 9,000 sq. ft.

27. The city will increase the FAR to 3.7 if the lot coverage is reduced to 50 percent, in order to obtain additional open space. If the first two floors have the maximum allowable area, the total floor area of the tower above the second floor would be increased by what percent?

   A. 33   **C.** 50

   B. 39   **D.** 100

28. A soil report discloses that a site is underlain by soft, loose soils. Which of the following is the best course of action for the architect to take?

   A. Notify the owner that the site may be developed only for recreation or other open space uses.

   B. Specify that all unsuitable soil be compacted in place.

   C. Obtain a recommendation from the soils engineer for the most feasible type of foundation system for the planned development.

   D. Notify the owner that any building foundations must extend through soft material and into soil of adequate strength.

29. Which of the following circulation patterns best describes the typical "shoestring" business district often called Main Street in many small American towns?

   A. Loop   **C.** Linear

   B. Radial   **D.** Spiral

30. Which of the following statements are correct concerning the microclimate of a city, as compared with that of rural areas in the same region?

   I.   The city's microclimate is warmer and drier.

   II.   The city has less rain, clouds, and fog.

   III. Wind velocities in the city are lower.

   IV. The city has less solar radiation.

   A. I, III, and IV   **C.** II and IV

   B. II, III, and IV   **D.** I and III

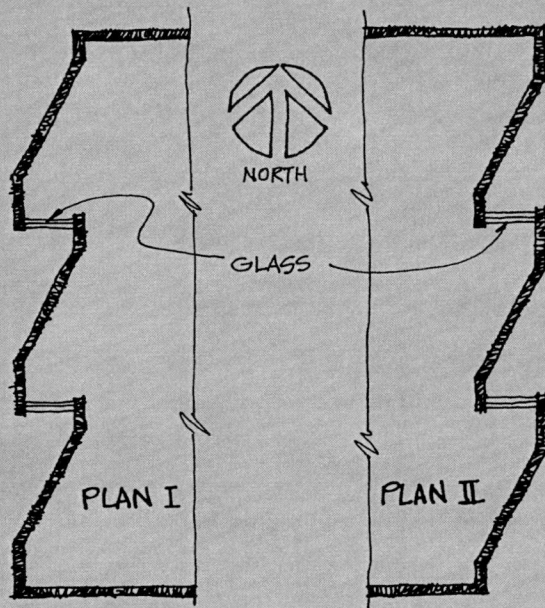

PLAN I   PLAN II

31. Two different sawtooth plans are shown above. Considering energy conservation, which statement is most correct?

A. Plan I is suitable for all climates.

B. Plan II is suitable for all climates.

C. Plan I is suitable for southern climates, and plan II is suitable for northern climates.

D. Plan II is most appropriate in southern climates, while plan I is best for northern climates.

32. A FAR bonus is usually the result of

A. flexible zoning.

B. a conditional use permit.

C. spot zoning.

D. density controls.

33. The community of Radburn became known as "The Town for the Motor Age" because it was

A. planned with full consideration for the automobile.

B. designed with streets wide enough to permit comfortable two-way vehicular circulation.

C. the first community to provide garages.

D. the first community designed for commuters.

34. Designing buildings to conserve energy has assumed great importance in recent years. In this regard, which of the following statements is INCORRECT?

A. A rectangular building absorbs less solar heat in the summer and more in the winter, if its long axis is in an east-west, rather than a north-south, direction.

B. A triangular building, with its base facing north and its apex facing south, is energy-efficient in a hot climate.

C. A building on a north-facing slope experiences heavier heating loads in winter, but lighter air conditioning loads in summer, as compared to buildings on other slopes.

D. In a cold climate, a small amount of glazing in the north wall is preferable to no glass at all, because it would allow heat gain from diffused radiation.

35. Select the most correct statement.

   A. Vegetation introduced into a site should generally be indigenous.

   B. Outstanding natural features, such as rock outcroppings and bodies of water, should be preserved if they can be justified economically.

   C. The ability of soil on a site to support vegetation is not generally a critical factor in site selection.

   D. A site located in a flood plain is not permitted to be developed for any purpose other than open space uses, such as recreation.

36. The City Beautiful Movement, from around the turn of the century, did all of the following EXCEPT

   A. launch a classic revival movement throughout the country.

   B. restore a human scale to the industrial cities.

   C. lead to an awareness of city planning.

   D. result in a profusion of dome-topped civic buildings.

37. Which of the following is the LEAST appropriate use?

   A. A library in a downtown urban center

   B. A residential development in hilly terrain

   C. A motion picture theater on the shore of a recreational lake

   D. A hotel adjacent to an international airport

38. High altitude sites, compared with those at sea level at the same latitude, have

   A. lower temperatures and more solar radiation.

   B. lower temperatures and less solar radiation.

   C. higher temperatures and more solar radiation.

   D. lower temperatures and the same amount of solar radiation.

39. In order to minimize foundation costs for buildings and other structures, developments should NOT be located

   A. in areas of noncompressible soil.

   B. on gentle slopes near bodies of water.

   C. on land comprising inorganic soils.

   D. where the site is composed of bedrock.

40. The greatest erosion tends to occur

   I. in sandy soils.

   II. in clayey soils.

   III. at the bottom of slopes.

   IV. on slopes.

   A. I and III   **C.** II and III

   B. I and IV   **D.** II and IV

41. In site planning, we are most interested in soil classification systems based on

   A. bearing capacity.

   B. drainage.

   C. erodibility.

   D. particle size.

42. The most common subsurface exploration consists of
    A. test borings.
    B. test pits.
    C. pile load tests.
    D. plate load tests.

43. If we create a water body in a natural setting, its shape should generally
    I. be natural.
    II. be geometric.
    III. be curvilinear.
    IV. appear man-made.
    A. I and II   C. II and IV
    B. I and III   D. III and IV

44. What is the name of the process by which water vapor escapes into the atmosphere from plants?
    A. Evaporation
    B. Lithification
    C. Transpiration
    D. Precipitation

45. A 50-year flood has what likelihood of occurring in any given year?
    A. 1 percent
    B. 2 percent
    C. 5 percent
    D. 10 percent

46. Which of the following are considered natural constraints to development of land?
    I. Unsightly vegetation
    II. Damp peat bogs
    III. Poisonous plants
    IV. Annoying insects
    V. Hostile neighbors
    A. I, II, and III   C. II, IV, and V
    B. II, III, and IV   D. III, IV, and V

47. Which of the following types of lamps provides the best color rendition?
    A. Mercury vapor
    B. Metal halide
    C. High pressure sodium
    D. Fluorescent

48. Which of the following types of lamps is the LEAST efficient, that is, produces the least lumens per watt?
    A. Incandescent
    B. Fluorescent
    C. Mercury vapor
    D. Metal halide

49. Large parking areas require nighttime illumination of no less than
    A. one footcandle.
    B. three footcandles.
    C. five footcandles.
    D. ten footcandles.

50. As the height of lighting standards increases, they must
    A. be spaced closer.
    B. be spaced further apart.
    C. have higher wattage lamps.
    D. use metal halide lamps.

51. An architect wants to create an earth berm that is 32 feet long and 4.5 feet high. The slope of the earth may not exceed 2-1/2:1. What will be the overall width of this berm?

    A. 11.25 feet   **C.** 22.5 feet

    B. 12.8 feet   **D.** 50 feet

The following data applies to question 52:

Building Site - 80' wide × 140' deep

Setbacks - Front - 25'

Side - 7.5'

Rear - 15'

Floor Area Ratio (FAR) - 0.85

Allowable Lot Coverage - 75 percent max.

52. What is the maximum floor area that would be permitted for a one-story building on this site?

    A. 6,500 square feet

    B. 8,400 square feet

    C. 9,520 square feet

    D. 11,200 square feet

53. As part of the drainage system of a large development, the architect has proposed using a retention pond that is 190 feet wide and 330 feet long. What will be the average depth of this pond in order for it to have a capacity of 5 acre-feet of water?

    A. 1.5 feet   **C.** 3.5 feet

    B. 2.75 feet   **D.** 7.35 feet

54. On a hillside lot, an architect wishes to place a garage as close as possible to the street. The garage floor has been established at elevation 172.0, the top of the six-inch high curb at the street is at elevation 168.5, and the maximum driveway slope allowed by ordinance is eight percent. Under these conditions, how close to the street can the garage be placed?

    A. 28 feet   **C.** 42 feet

    B. 32 feet   **D.** 50 feet

55. A shopping center parking area must accommodate 320 cars. What is the approximate amount of undeveloped flat land that should be acquired by the developers for this parking area?

    A. 2.6 acres   **C.** 4.2 acres

    B. 3.8 acres   **D.** 5.4 acres

56. A local zoning ordinance permits an eave to project 18 inches into a required side yard. If a building has a 7-foot overhanging eave and the side yard setback is six feet, how close to the side property line can the building be placed?

    A. 13.0 feet   **C.** 6.0 feet

    B. 11.5 feet   **D.** 4.5 feet

The following data applies to questions 57 and 58:

Building Site - 120' × 184'

Floor Area Ratio (FAR) - 5

Efficiency Ratio - 70 percent

Setback - 16' from front property line

Utility Easement - 8' from rear property line

Open Space Bonus - four square feet of gross floor area for each square foot of open space

57. What is the minimum number of stories above grade for a building on this site, if it is developed to the maximum allowable square footage? The project includes an open entrance forecourt, behind the front setback, which measures 60' × 80'.

   A. Six   **C.** Eight

   B. Seven   **D.** Nine

58. Using the data in the previous question, what is the net floor area of the largest building that can be built on this site? Disregard the open space bonus.

   A. 50,400 square feet

   B. 77,280 square feet

   C. 96,000 square feet

   D. 110,400 square feet

Questions 59 and 60 are based on the zoning summary shown on the following page.

59. A 72-unit apartment project is planned, which will comprise three individual buildings with 24 two-story units in each building. Which of the following land parcels will NOT provide sufficient acreage for the proposed development?

   A. Five acres in zone RM 16

   B. Five acres in zone RM 16-1

   C. Three acres in zone RM 32

   D. Two acres in zone RM 48

60. Referring to the same zoning regulations, what is the maximum size five-story building that can be built on a corner lot in a RM 48 zone, if the lot has a frontage of 90 feet on one street and 160 feet on the other?

   A. 7,800 square feet per floor

   B. 8,450 square feet per floor

   C. 8,640 square feet per floor

   D. 9,100 square feet per floor

61. The planning phase of a sustainably designed architectural project should include which of the following elements?

   I.   Native landscaping that is aesthetically pleasing and functional.

   II. Designing structures in the floodplain that can resist the forces of flood waters.

   III. Consideration of sun orientation, topographic relief, and the scale of adjacent buildings.

   IV. Locating projects within existing neighborhoods that are adjacent to public transportation.

   A. I, II

   B. I, III

   C. I, III, IV

   D. all of the above

Summary: Zoning Regulations

ZONE	LAND USE	MAX. HT.	REQUIRED YARDS			MINIMUM LOT AREA (SQ.FT.)		MIN. LOT WIDTH	MAX. BLDG. AREA	PARKING SPACES PER D.U.
			Front	Side	Rear	Per Lot	Per D.U.			
RS 1	ONE FAMILY RESIDENCE 1 unit/acre	32'	25'	5' min. or 10% of lot width, 10' max.	25'	40,000	40,000	100'	35%	2 covered
RS 2	ONE FAMILY RESIDENCE 2 units/acre Any RS-1 use	32'	25'	5' Same as RS-1	25'	20,000	20,000	100'	35%	2 covered
RS 4	ONE FAMILY RESIDENCE 4 units/acre Any RS-1 use	32'	25'	5' Same as RS-1	25'	12,000	12,000	75'	35%	2 covered
RS 6	ONE FAMILY RESIDENCE 6 units/acre Any RS-1 use	32'	25'	5' Same as RS-1	25'	7,200	7,200	55'	35%	2 covered
RM 16	MULTI-FAMILY RESIDENCE 16 units/acre Any RS use	36'	20'	10' corner lots require 15'	10'	7,200	2,750	55'	40%	2 covered per unit 550 sq.ft. or larger
RM 16-1	MULTI-FAMILY RESIDENCE 14 units/acre Any RS use	36'	40'	10' corner lots require 30'	20'	12,000	3,000	75'	35%	Same as RM-16
RM 32	MULTI-FAMILY RESIDENCE 32 units/acre Any RS use Group residential	36'	20'	10' Corner lots require 15'	10'	10,000	1,360	60'	60%	Same as RM-16
RM 48	MULTI-FAMILY RESIDENCE 48 units/acre Any RM-32 use	60'	20'	10' Corner lots require 15'	10'	10,300	910	60'	60%	Same as RM-16

The examination answers and explanations will be found on the following pages.

Do not look at the answers until you have completed the exam.

EXAMINATION ANSWERS

1. **C** The slope at point A is represented by equally-spaced parallel contour lines, which indicates that the land rises or falls an equal amount between each contour. In other words, the slope is uniform, and a section cut perpendicular to those contours would show a straight line profile.

2. **B** The situation at point B appears unusual, since the 21 contour overlaps the 19 contour. This can only happen where land of a higher elevation hangs over the land below, such as a cave or an overhanging cliff.

3. **C** Water always flows in a direction perpendicular to the contours. Thus, at point C water would flow in a southerly direction, from higher elevations to lower.

4. **A** The landform at point D is represented by a contour (19) that closes on itself. Therefore, it can only be either a summit or a depression. Because the adjacent contour (20) is higher, the configuration is a depression.

5. **D** Point E is in an area relatively free of contour lines. Thus, there is very little change in elevation, and it appears to be the flattest area on the map, since no other area is as wide open.

6. **B** The slope at point F is represented by roughly parallel contours which are spaced at increasing intervals as they move uphill. Therefore, the slope is convex. Convex slopes always have closer contours at the lower elevations.

7. **A** Contours that point in the downhill direction, as seen at point G, represent ridges. If the contours pointed uphill, the configuration would be a valley.

8. **B** The gradient, or steepness, of a road is represented by the formula $G = V/H$, where G is the gradient in percentage, V is the vertical rise, and H is the horizontal distance in which the rise takes place. To find the vertical rise, we transpose the formula to read $V = GH$. Thus, $V = .05 \times 150 = 7.5$ feet, which is the total difference in elevation over the 150-foot distance. Downhill from the point of 132.5 elevation, therefore, the elevation will be $132.5 - 7.5$ or 125 (correct answer B).

9. **A** A one-way system will usually speed the flow of traffic because the danger from oncoming traffic, especially at intersections, is eliminated (correct answer A). The direction of traffic flow does not affect the distance from utility mains to connection points (B). A one-way system has at least as many intersections, to allow for changes in direction of travel (C). Pedestrians must cross traffic arteries regardless of directional flow unless vertically separated from vehicles (D).

10. **B** Irregular and steeply sloping land requires many changes in the amount and direction of slope of utility lines. This is particularly costly because of the need for manholes and pumping stations for sewer lines that depend on the natural slope to drain. A dead flat site, however, can be equally troublesome because sewers and storm drains must slope to connection points, which may require pumping to reach the invert elevation of mains in streets (incorrect answer A). Excavation costs, length of runs, and changes in direction all add to the cost of installation (correct answer B, incorrect answer C). Changes in the horizontal direction occur where street layouts have sharp curves and bends, not necessarily where the terrain is steep (D is incorrect).

11. **C** Medieval towns were often fortified because of frequent rivalries and wars among feudal lords. When a town was established, around a church, monastery, or castle, protective walls were constructed for safety and security. And because labor and materials were limited, the walls were

constructed in a circular fashion, thus circumscribing the greatest area with the least amount of enclosure (correct answer C). Most towns continued to grow within the confines of these walls, but when expansion became necessary, it was accompanied by additional encircling walls. None of the other choices has much validity; straight walls could be built on irregular terrain (A), the technology for right angles was available (B), and finally, most towns developed and expanded after the walls were built (D).

12. **B**   All of the devices mentioned are typical of construction in the hot-arid zone (correct answer B). In this zone, there is a wide daily variation of temperature, and thick walls, especially of masonry or adobe, tend to make interior spaces cooler in the daytime and warmer at night by absorbing the sun's heat and radiating it to interior spaces at night. Wide overhangs block the high summer sun, high ceilings allow air circulation, and limited fenestration reduces direct and reflected radiation heat gain.

13. **B**   By referring to the sun chart on page 7, one can see that in the winter, the sun is low and oriented mostly to the south. Terrace A, located to the north and east, would therefore be cold in the winter, since it would receive little solar radiation then. However, the location of terrace A is ideal in the summer (correct answer B), since the north side is cool and the east side receives sun only in the morning. C and D are incorrect: terrace A receives more morning sun in the summer and very little afternoon sun at any time.

14. **B**   Trees C, in the northwest, provide protection from the northwest winter winds if they are evergreen (II). They also shade the low afternoon sun of summer, which is to the west and northwest, according to the sun chart (III). The correct answer is therefore B.

15. **C**   Only trees C should be evergreen, in order to block the winter winds. Trees D and E should be deciduous, to allow the winter sun to enter. It is true that trees D tend to block the summer breezes from the southwest, but this would be true whether they are evergreen or deciduous.

16. **D**   Grading costs can be reduced in several ways, including balancing cut and fill (A), so that soil will not have to be hauled away or imported. Avoiding winter work (B) saves the cost of operating in wet, muddy, or frozen conditions, all of which slow down equipment. If the land covered by construction is reduced (C), perhaps by adding another story, parking under buildings, etc., grading work will also be reduced. Finally, and unfortunately, it is usually cheaper to remove trees than to save them (correct answer D), simply because preserved trees become an obstacle to the grading process and require special protection and care. However, the ultimate cost to replace mature trees may justify spending more for additional grading.

17. **A**   Open storm drainage systems are those on the surface of the ground, such as gutters, swales, sloped planes, etc. Closed drainage systems occur below ground level and include various collector drains and closed pipes. Open systems are not only less expensive to install (correct answer A), but they are also easier to maintain. However, they do not necessarily have greater capacity (B), nor are their gradients more flexible. In fact, the gradients of subsurface systems are generally more flexible, because pipes can be sloped independent of the actual ground slope. Finally, a closed drainage system is preferable for eliminating standing water, since an area drain and underground pipe

can readily accomplish what the natural topography may be unable to do.

18. **C** Dashed, not solid, contour lines are used to represent the existing or natural topography (III is incorrect). Solid contour lines, on the other hand, represent proposed modifications to the existing landform (II). All contour lines, existing or revised, connect points of equal elevation (I), and although contour lines may close on themselves (IV is incorrect), indicating a summit or a depression, they may never split in two (V), since this would represent an impossible topographic configuration. The correct combination of answers is found in choice C.

19. **C** Solar radiation that reaches the earth is in two forms: direct radiation from the sun and diffuse radiation from the sky. Flat plate collectors absorb both direct and diffuse radiation (A is incorrect). B is also incorrect: collectors should be oriented to the south, and a variation more than 10° to 15° will decrease the efficiency of the collector and thus require more collector area. Normally, it is not economically feasible to supply 100 percent of the hot water requirement by solar heating, except in hot areas (incorrect statement D). C is correct; water is inexpensive and efficient and the most common heat transfer and storage medium used for solar systems.

20. **A** The thickest layer of topsoil is found near the toe of the slope as a result of the deposition of organic matter eroded from the slope. Consequently, that is where the vegetation tends to be thickest. A is therefore the correct answer. The greatest erosion occurs at the steepest part of the slope, which is usually at midslope. Therefore, the topsoil there is thin and the vegetation sparse (B is incorrect).

21. **D** Don't be confused by the question; this is actually a simple problem. The only fact to consider is this: Since the east end of the wall terminates at about elevation 11, then the wall must extend to that elevation in order to retain the earth at the south side of the wall, regardless of the level area's elevation. The east end of the wall would actually be exposed seven feet above the adjacent grade at the north (11 − 4 = 7), but the elevation at the wall's top would be 11 feet, or perhaps a few inches higher (correct answer D).

For questions 22 through 27, refer to the sketches on page 193.

22. **C** The maximum lot area that may be covered by a building is the total area of the site multiplied by the maximum lot coverage. The site area is 20,000 s.f. (100 × 200) and the maximum lot coverage is 60 percent. The maximum lot area that may be covered by a building is therefore 20,000 × 0.60 = 12,000 s.f.

23. **A** The FAR is the ratio of total allowable floor area to total lot area. Since the lot area is 20,000 s.f. and the FAR is 3, the maximum allowable building floor area is the FAR times the lot area = 3 × 20,000 = 60,000 s.f.

24. **B** The parking requirement is 400 square feet for every 1,000 square feet of building, area. The 60,000 s.f. building will require 24,000 s.f. of parking [(60,000 ÷ 1,000) × 400 = 24,000 s.f.]. At 12,000 s.f. per parking level (same area as ground floor), 24,000 s.f. of parking will require two levels (24,000 ÷ 12,000 = 2).

25. **C** The maximum allowable lot coverage yields 12,000 s.f. at both ground and second floor levels for a total of 24,000 s.f. (12,000 × 2 = 24,000 s.f.). Setbacks reduce the maximum area of floors above the second floor to 9,000 square feet per floor

[(200 − 25 − 25) × (100 − 20 − 20)] = 150 × 60 = 9,000 s.f. Because the total building area is 60,000 s.f. and the first two floors total 24,000 s.f. (12,000 × 2 = 24,000 s.f.), the remaining allowable area is 60,000 − 24,000 = 36,000 s.f. 36,000 s.f. divided by the maximum area per floor of 9,000 s.f. equals 4. Two floors × 12,000 s.f. plus four floors × 9,000 s.f. equals a total of six floors. This is the fewest number of floors able to accommodate the maximum allowable area of 60,000 s.f.

26. **B** Based on a total area of 24,000 s.f. for the first two floors of the building and the fact that the remaining area of 36,000 s.f. (60,000 − 24,000 = 36,000 s.f.) is to be distributed equally over the remaining nine floors (11 floors − 2 floors = 9 floors), the maximum area per floor will be 36,000 s.f. divided by 9 floors = 4,000 s.f. per floor.

27. **C** An increase in the FAR to 3.7 will increase the total allowable building area from 60,000 s.f. (20,000 × 3 = 60,000), to 74,000 s.f. (20,000 × 3.7 = 74,000). At the same time, a reduction in lot coverage from 60 percent (20,000 × .6 = 12,000 s.f.) to 50 percent (20,000 × .5 = 10,000 s.f.) will reduce the allowable area of the first two floors from 12,000 × 2 (24,000 s.f.) to 10,000 × 2 (20,000 s.f.). Consequently, the remaining allowable floor area above the second floor will be 54,000 s.f. (74,000 s.f. − 20,000 s.f.), an increase of 18,000 s.f. (54,000 − 36,000) or 50 percent (18,000 ÷ 36,000 = 0.5).

28. **C** The information given in this question is rather skimpy. We don't know the depth of the underlying soft soil or its suitability for compaction, and therefore we don't choose answer B. We are not sure that building foundations extending through the soft soils provide the best solution (D is questionable). And knowing as little as we do, we certainly can't limit the site development to recreation (incorrect answer A). The only reasonable answer is C: the architect should find out what kind of foundation system is most appropriate for the development being planned.

29. **C** A loop system (A) closes upon itself and improves the flow characteristics by offering two choices of direction to each destination. In a radial system (D) all channels of circulation spread out from the center. A linear pattern of circulation (correct answer C) consists of a single line or a parallel series, to which all origins and destinations are directly attached. A spiral configuration (D) is a single, continuous path that originates from a central point, revolves around it, and becomes increasingly distant from the point of origin.

30. **A** Wherever man intrudes into nature, he modifies the microclimate. This is particularly true in the city, where an entirely new microclimate is created, as a result of the elimination of natural ground cover and the emission of heat. The city is warmer and drier (I), with more rain, clouds, and fog (II is incorrect), lower wind velocities (III), and less solar radiation (IV) because of clouds and haze. I, III, and IV are true statements, making A the correct answer.

31. **D** In northern (cold) climates, there should be minimal glass on the north wall to minimize heat loss, and maximum glass on the south wall to maximize winter heat gain. This can be accomplished by the sawtooth design shown in plan I. In the warmer, southern climates, however, the situation is different, and energy conservation in such areas is directed toward preventing heat gain. Glazing in the north wall minimizes heat gain, since the north side receives minimal solar radiation. Sawtooth pattern II is therefore appropriate for southern climates, making D the correct answer.

9 FLOORS @ 4000 S.F. = 36,000 S.F.

OR

6 FLOORS @ 6000 S.F. = 36,000 S.F.

OR

4 FLOORS @ 9000 S.F. = 36,000 S.F.

2 FLOORS @ 12,000 S.F. = 24,000 S.F.

HEIGHT LIMIT - 11 FLOORS

SETBACK

GROUND FLOOR

PARKING

TOTAL BUILDING = 60,000 S.F.

2 LEVELS @ 12,000 S.F. = 24,000 S.F.

LOT COVERAGE

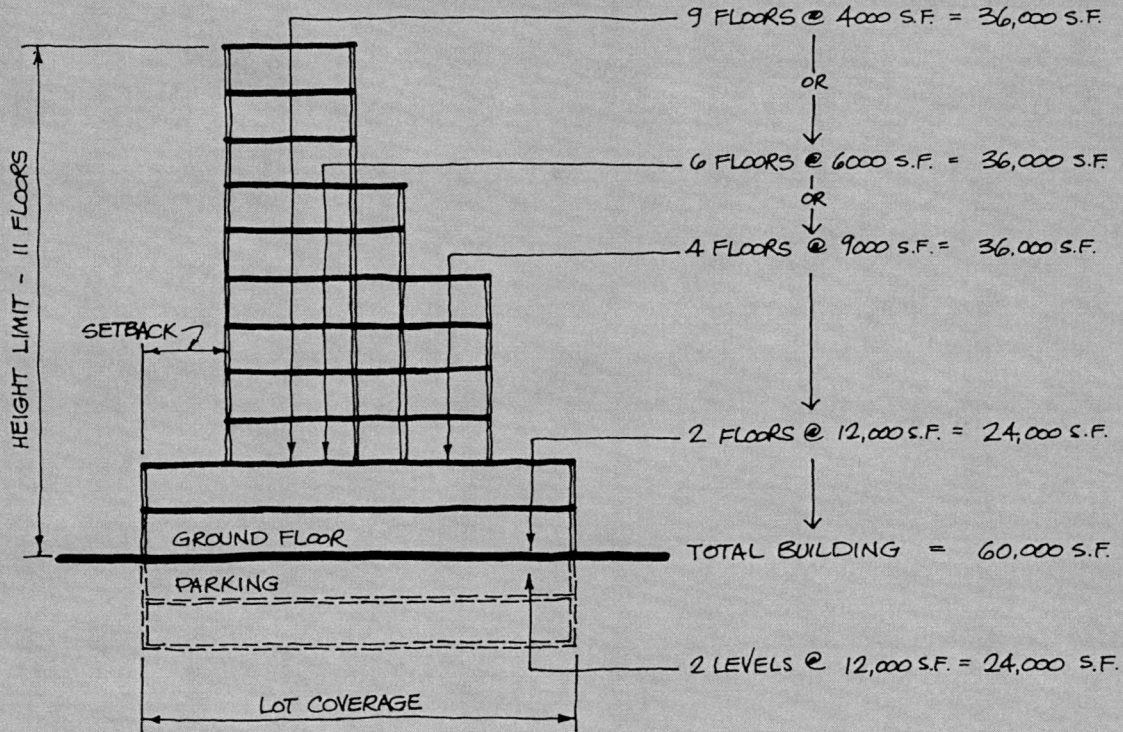

TOTAL SITE AREA = 100' x 200' = 20,000 S.F.
F.A.R. = 3 ; TOTAL ALLOWABLE BUILDING AREA = 20,000 S.F. x 3 = 60,000 S.F.
LOT COVERAGE = 60% x 20,000 S.F. = 12,000 S.F.

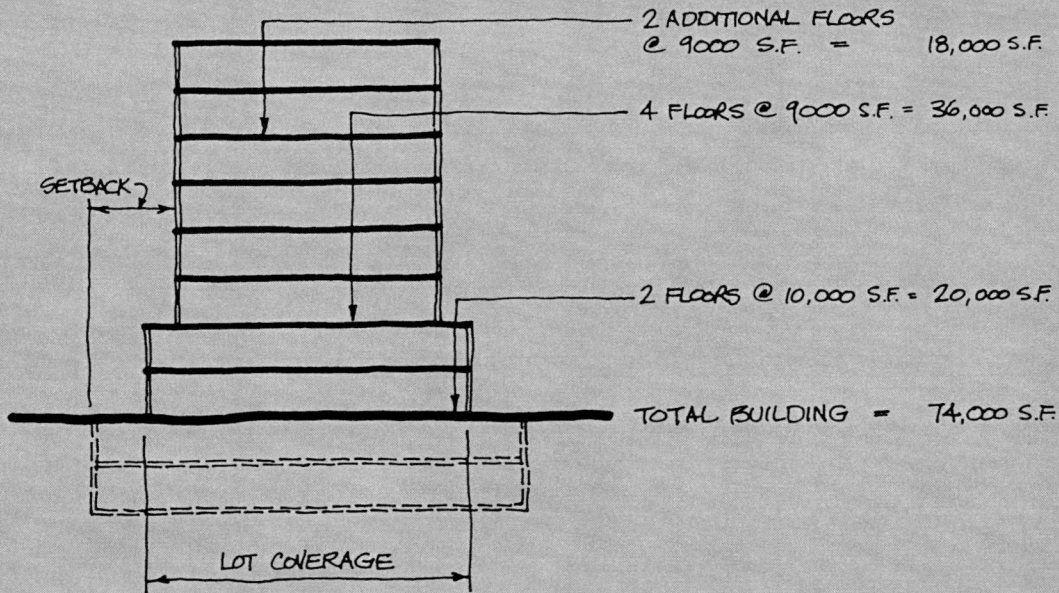

2 ADDITIONAL FLOORS
@ 9000 S.F. =   18,000 S.F.

4 FLOORS @ 9000 S.F. = 36,000 S.F.

SETBACK

2 FLOORS @ 10,000 S.F. = 20,000 S.F.

TOTAL BUILDING = 74,000 S.F.

LOT COVERAGE

TOTAL SITE AREA = 100' x 200' = 20,000 S.F.
F.A.R. = 3.7 ; TOTAL ALLOWABLE BUILDING AREA = 20,000 S.F. x 3.7 = 74,000 S.F.
LOT COVERAGE = 50% x 20,000 S.F. = 10,000 S.F.

32. A   Flexible zoning attempts to overcome some of the rigidities of traditional zoning, in order to be relevant to changing patterns of development. An increase in the floor area ratio is usually granted as an incentive for the developer to provide certain amenities that would not normally be required under the provisions of the applicable zoning ordinance; hence the phrase incentive zoning. The latter, as well as the planned unit development, the floating zone, and contract zoning are all types of flexible zoning.

33. A   Radburn, in New Jersey, was designed by Henry Wright and Clarence Stein to demonstrate their concept of the super-block. In their innovative plan, traffic streets surrounded, rather than crossed, the living areas. Housing units were grouped around cul-de-sac roads, which became access/service roads, and these units were reversed so that service areas faced the roads, while living areas faced generous rear gardens. Pedestrian walks were separated from vehicular traffic, often with underpasses. In all, Radburn was planned with full consideration for the automobile (correct answer A), as well as for the residents of this small community. The remaining choices all have some grain of truth, but are less inclusive than answer A.

34. D   Let's examine each statement. A is correct: the south wall receives maximum solar radiation in the winter and minimum in the summer. East and west walls, on the other hand, receive maximum radiation during the summer and minimum during the winter months. Orienting a building east-west, therefore, results in less solar heat gain in the summer and more in the winter. A triangular building facing north (B) elimi-nates the south wall, where heat gain is great, and is therefore energy-efficient in a hot climate. C is also correct: north-facing slopes are always cooler than other slopes, because they receive the least amount of solar radiation. A building on a northerly slope therefore requires more heating in the winter, but less cooling in the summer. D is incorrect and therefore the answer to this question. In a cold climate, there should be minimal glass in the north wall, since the greatest heat loss occurs there, and having no glass at all in the north wall would reduce energy consumption even more.

35. C   This is a question in which each of the four choices has a grain of truth to it, and we must decide which of them is the most true. While much of the vegetation intro-duced may be indigenous, it is not necessarily so, and we rule out choice A. Outstanding natural features should be preserved if at all possible—the economics of such preservation are of secondary concern (incorrect choice B). While open space uses are preferred in flood plain sites, other uses are not always prohibited (D is incorrect). Choice C is the correct answer: a soil's fertility is not generally critical in site selection, since deficiencies in plant-growing qualities can frequently be remedied.

36. B   The City Beautiful Movement resulted from the enormous influence of the Chicago World's Fair of 1893. On the shores of Lake Michigan, Daniel Burnham created the big, spacious, and classically beautiful "White City," which launched a classic revival that swept the country (A). It resulted in a profusion of civic centers (D), grand plazas, broad avenues, and the obligatory monuments and fountains. This surge of development led to the creation of city planning organizations (C), but the scale of development, far from restoring the human element (correct answer B), became monumental, grandiose, and lavish.

37. C The least appropriate use is a theater at the lake shore (correct answer C). Motion picture theaters are windowless buildings, while most recreational lakes provide attractive views. Therefore, if a theater near a lake is desirable, it should preferably be placed well away from the visually valuable water's edge. All of the other combinations are not only appropriate, but prevalent throughout the country.

38. A As the elevation increases, the temperature decreases (about 3°F for every 1,000 feet), because the thinner air at higher altitudes cannot hold as much heat. This eliminates choice C. But what about solar radiation—how is it affected by elevation? As the altitude increases, the amount of atmosphere that solar radiation must pass through decreases. Therefore there is more solar radiation at high altitude sites, making A the correct answer.

39. D Where the land is mostly made up of bedrock, the soil cover is either thin or absent, there is very little vegetation, and excavation for footings is more difficult and hence, more expensive, than in areas where the soil is well drained, inorganic, with a cover of several feet.

40. B Sandy soils tend to be eroded more than clayey or gravelly soils (I is correct, II is incorrect), and erosion is greater at the steepest part of a slope than at the bottom of the slope, where eroded material is deposited (IV is correct, III is incorrect).

41. D While there are many soil classification systems, in site planning we are most interested in those based on particle size, since that largely determines the properties of a soil, such as drainage, bearing capacity, and erodibility.

42. A Subsurface explorations are often necessary, in order to obtain information about the soil, from which its bearing capacity and other properties may be determined. The most common subsurface exploration consists of a series of exploratory borings.

43. B A natural water body is usually curvilinear and free-flowing in shape. If we create a water body in a natural setting, we usually attempt to emulate nature by making the pond or lake curvilinear.

44. C A great deal of precipitation is transpired into the atmosphere, through the action of plants.

45. B The term 50-year flood refers to a flood of a magnitude such that it is likely to occur only once every 50 years; therefore, the likelihood of its occurrence in any given year is 2 percent.

46. B Poisonous plants (III) or annoying insects (IV) may modify development, while peat bogs (II) may cause a developer to abandon the site completely. Unsightly vegetation (I), whatever that might imply, is rarely a cause for concern, since such features are easy to modify or remove. Hostile neighbors (V), however, are not so easily modified or removed. On the other hand, they cannot be considered a natural development constraint. Therefore, the correct combination of answers is found in choice B.

47. D Fluorescent lamps provide better overall color rendition than high pressure sodium, metal halide, or mercury vapor.

48. A Incandescent lamps are the least efficient lamps because they consume a great deal of energy as heat.

49. A The average intensity for lighting large parking areas is one footcandle. See page 109.

50 C The taller the lighting standard, the better the coverage; however, higher wattage lamps must be used in order to provide the same intensity as lamps mounted closer to the surface illuminated.

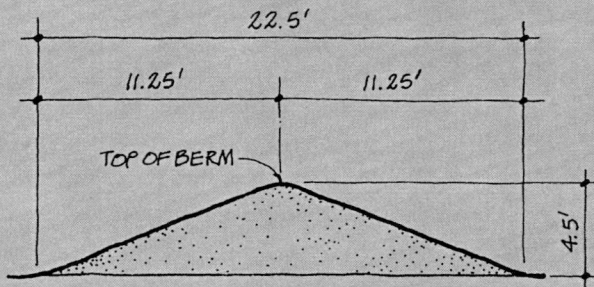

51. C In this problem, one must recognize that the length of the berm is irrelevant. What concerns us is only the berm height and slope. With a height of 4.5 feet and a slope of 2 – 1/2:1, the horizontal dimension H is determined from the following ratio: 2.5/1 = H/4.5. H is therefore 4.5 times 2.5 = 11.25 feet. This dimension is half of the overall berm width. The overall width is twice that distance, or 22.5 feet (correct answer C), as shown in the sketch above. Remember that when slopes are expressed as ratios, the larger number is the horizontal dimension. Thus, a ratio of 2-1/2:1 indicates a horizontal run of 2.5 feet for every foot of vertical rise.

52. A In this type of problem, which is not unusual on the actual exam, we must calculate the maximum permitted floor area based on each of the given constraints. The most restrictive of these constraints governs—that is, the one which results in the smallest floor area. We begin with the allowable lot coverage, which is given as 75 percent. This factor is applied to the *gross* area of the site (ignoring the setback data), which is 80 × 140 = 11,200 square feet. Thus, .75 × 11,200 = 8,400 square feet of floor area. Next, we compute the allowable floor area using the floor area ratio (FAR)

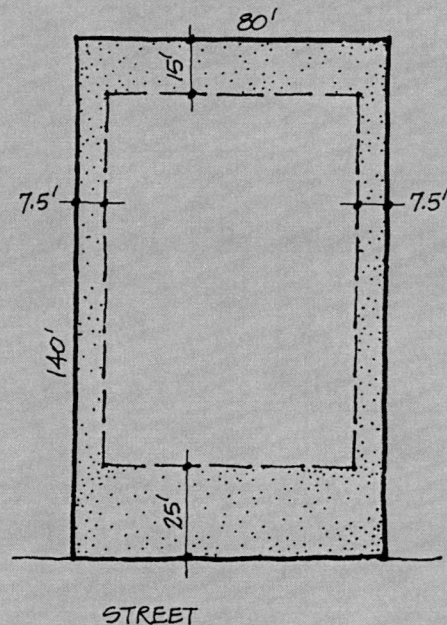

STREET

of 0.85. This factor is also applied to the *gross* site area, so we have 11,200 × 0.85 = 9,520 square feet. The final constraint is imposed by the setbacks. The net area after subtracting the setbacks is calculated as follows: [80 – (2 × 7.5)] × [140 – (25 + 15)] = 65 × 100 = 6,500 square feet. Thus, this constraint governs, since it is the most limiting of all three restrictions, and answer A (6,500 square feet) is correct. Incidentally, if this were not specified as a one-story building, the floor area ratio would govern, since the area in excess of 6,500 square feet could be located on another story. In other words, since the FAR applies to the *total* building area, irrespective of the number of stories, the 9,520 square feet could be divided among two, three, or even 10 stories, so long as the ground floor coverage did not exceed 6,500 square feet.

53. C To solve this problem, one must know that an acre-foot of water is the volume of an acre of water that is one foot deep, and that an acre is 43,560 square feet in area.

Because the pond's capacity is five acre-feet, the volume is $43,560 \times 5 = 217,800$ cubic feet of water. The area of the retention pond is $190 \times 330 = 62,700$ square feet. The depth is the volume divided by the area $= 217,800 \div 62,700 = 3.47$ feet. Because the closest available answer is 3.5 feet, the correct answer is choice C.

54. D An 8 percent slope means that the driveway rises eight feet vertically for every 100 feet of horizontal distance, or one foot vertically for every 12.5 feet of horizontal distance. Remember the formula $G = V/H$, where G is the gradient in percent, V is the vertical rise, and H is the horizontal run. The curb elevation is given as 168.5; however, the *street* elevation is six inches lower than the top of the curb, or 168.0. Therefore, the difference in elevation between the garage and the street is $172 - 168$ or four feet. Thus, with an eight percent slope and a vertical distance of four feet, the horizontal distance from the street to the garage will be $H = V/G = 4/.08 = 50$ feet (correct answer D).

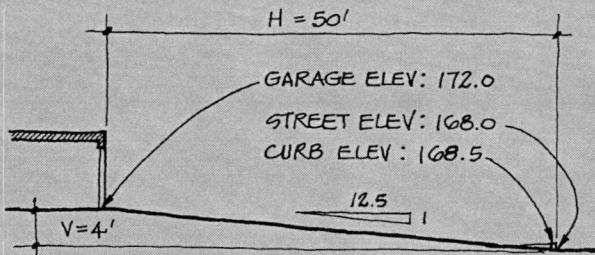

55. A The approximate overall area required for parking, including stalls, aisles, drives, etc. varies between 300 and 400 square feet per vehicle. One must also know that there are 43,560 square feet in an acre. Consequently, if one uses an average area of 350 square feet per car, the total area required for parking is 320 cars $\times$ 350 = 112,000 square feet. This total area must be converted into acres; thus, 112,000 $\div$

43,560 = 2.57 acres, and the closest choice is found in correct answer A. If you had used any other vehicle area requirement between 300 and 400 square feet per car, the most correct choice would still be answer A.

56. B For this kind of question, candidates should make a quick sketch, as shown below.

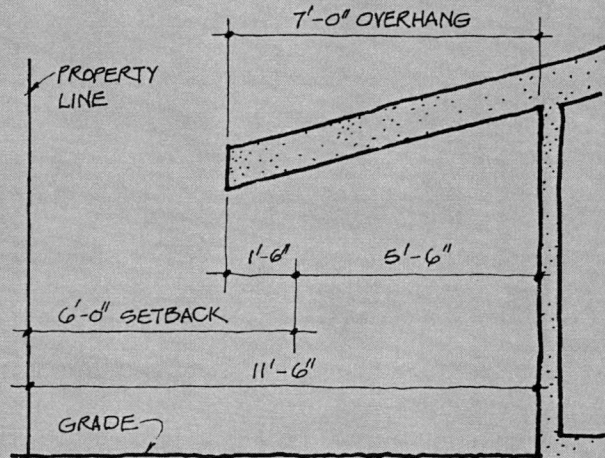

One can readily see that the building must be placed 5.5 feet away from the setback line (seven-foot overhang minus the 1.5-foot overhang projection). To this dimension we must add the dimension of the setback itself. Thus, $5.5 + 6.0 = 11.5$ feet, which is correct answer B.

57. D This question is about as complicated as a calculation-type problem can be. However, a rough sketch, as shown on the following page, will help to clarify it.

The site area is $120 \times 184 = 22,080$ square feet. Next, you multiply this gross site area by the floor area ratio: $22,080 \times 5 = 110,400$ square feet. To this figure must be added the open space bonus, which is the bonus awarded for *not* building on any buildable area of the site. The open forecourt is $60 \times 80 = 4,800$ square feet, and for each square

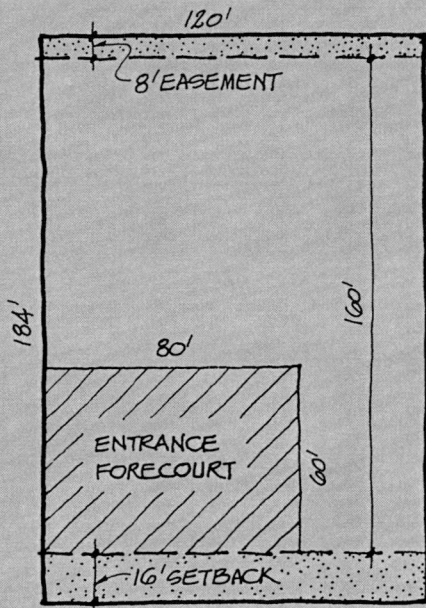

foot of open space we can add four additional square feet of gross floor area.

Thus, $4,800 \times 4 = 19,200$ square feet. The total allowable gross building area, therefore, is $110,400 + 19,200 = 129,600$ square feet. Now, we must figure the allowable site area on which the structure can be built. First, the front setback and rear easement must be subtracted from the total lot depth: $184 - (16 + 8) = 160$. That yields a net area of $120 \times 160 = 19,200$ square feet. Next, we must deduct the open entrance forecourt area of $60 \times 80 = 4,800$ square feet. Thus, $19,200 - 4,800 = 14,400$ square feet of area on each floor of the building. Finally, we divide the total allowable

building area by the allowable area on each floor in order to determine the number of stories. Thus, $129,600 \div 14,400 = 9$ stories, which is answer D. As you can see, each of the calculations is relatively simple, but the entire process is complex and time-consuming.

58. **B** Referring to the calculations for the previous question, the area of the site is 22,080 square feet. With an allowable floor area ratio of 5, the allowable gross building area is $5 \times 22,080 = 110,400$ square feet. The efficiency ratio is the ratio of net building area to gross building area, which is specified as 70 percent. In order to determine the maximum *net* floor area, therefore, we multiply the gross area by the efficiency ratio. Thus, $110,400 \times .70 = 77,280$ square feet, as found in correct answer B.

59. **B** This kind of question is not as difficult as it is tedious and time-consuming. To begin with, candidates must learn to recognize any extraneous information that has no bearing on the problem. For example, it is completely irrelevant that the units are each two-story and contained in three separate buildings. In reviewing the zoning summary, however, one can distinguish two distinct factors that might govern the total acreage required for this project. The first of these is the minimum lot area per dwelling unit (D.U.). Using the figures shown, one should construct a chart as follows:

Zone	Lot Area/ D.U.	No. of Units		Required Lot Area		SQ. FT./ Acre	Acres Req.	Acres (Actual)
RM 16	2,750 × 72		=	198,000	÷	43,560	= 4.55	5
RM 16-1	3,000 × 72		=	216,000	÷	43,560	= 4.96	5
RM 32	1,360 × 72		=	97,920	÷	43,560	= 2.25	3
RM 48	910 × 72		=	65,520	÷	43,560	= 1.50	2

Based on this chart, one can readily see that the acreage provided in each of the answers is greater than the acreage required. Therefore, since we are looking for a parcel with insufficient acreage, none of these choices can be correct. Next, we must check the maximum units per acre, found under the column labeled "Land Use" and construct another chart, as shown below.

Zone	Units/ Acre		Acres		Units Allowed
RM 16	16	×	5	=	80
RM 16-1	14	×	5	=	70
RM 32	32	×	3	=	96
RM 48	48	×	2	=	96

In this case, we see at once that the only possible answer is B, 5 acres in zone RM 16-1, because that is the only choice in which the maximum number of units permitted is less than the 72 units planned. If one had checked out the last factor first, a good deal of time might have been saved. However, it is generally impossible to read a question like this and know for certain which zoning restriction will govern the answer. In any event, don't let an involved question like this jeopardize your performance on the test; if you don't get the right answer in a reasonable length of time, move on.

60. B From the zoning regulations, the maximum building area allowed (lot coverage) is 60 percent of the lot area. The lot area is 90 × 160 = 14,400 square feet, and multiplying this by 60 percent, we obtain 8,640 square feet. Next, we make a sketch of the lot, as shown at the right, and indicate all the required setbacks specified in the zoning regulations. Note that the narrower dimension of a corner lot is

generally considered to be the front, and that the required side yard along the street of a corner lot is 15 feet, not the usual 10 feet. When the lot area is reduced by the required yards, as shown, the buildable area becomes: $[90 - (15 + 10)] \times [160 - (20 + 10)] = 65 \times 130 = 8,450$ square feet. Because the lot development is limited by the most restrictive regulation, the maximum allowable area per floor is governed by our second calculation, as shown in answer B. The five-story requirement is irrelevant; however, it would be permitted in this zone, as long as the maximum allowable height of 60 feet is not exceeded.

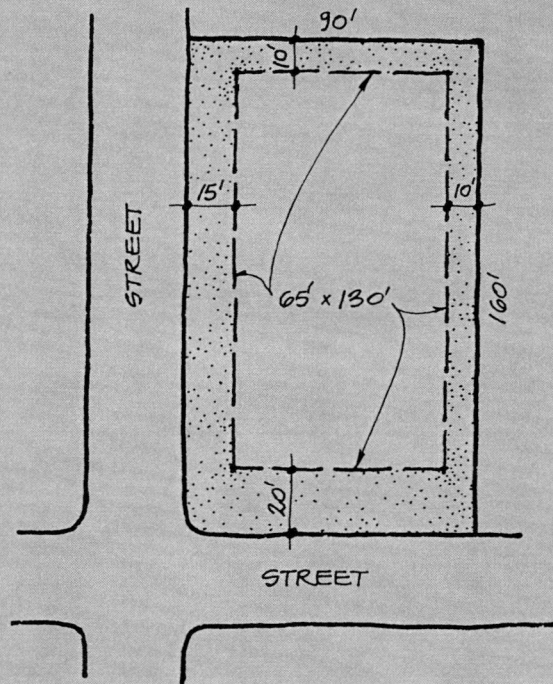

61. C I, III, IV

Choice I is correct. Designing with native landscaping is preferred to using exotic or imported plant types. Indigenous plants tend to survive longer, use less water, and cost less.

Choice II is not correct. Placing any structure in a floodplain, even those that resist floodwater, is not desirable. Placing buildings in a floodplain can increase flooding further down stream.

Choice III is also correct. Buildings sensitive to the benefits of solar orientation and passive and active solar gain techniques save energy and are more visually aligned with local climatic conditions.

Choice IV is correct as well. Infill development and proximity to a variety of transportation options are design principles that benefit the inhabitants and their environment.

INDEX